Environmental Science

Series editors: R. Allan U. Förstner W. Salomons

Springer

Berlin
Heidelberg
New York
Barcelona
Budapest
Hong Kong
London
Milan
Paris
Santa Clara
Singapore
Tokyo

Rudolf Reuther (Ed.)

Geochemical Approaches to Environmental Engineering of Metals

With 68 Figures and 44 Tables

Springer

Volume Editor

DR. RUDOLF REUTHER
Environmental Geochemistry & Consulting
Hindenburgstraße 17
D-79320 Albertshofen
Germany

ISBN-13:978-3-642-79527-5

Cataloging-in-Publication Data applied for

Die Deutsche Bibliothek – CIP-Einheitsaufnahme
Geochemical approaches to environmental engineering of metals / R. Reuther (ed.). – Berlin ; Heidelberg ; New York ; Barcelona ; Budapest ; Hong Kong ; London ; Milan; Paris ; Santa Clara ; Singapore ; Tokyo : Springer 1996
(Environmental science)
ISBN-13:978-3-642-79527-5 e-ISBN-13:978-3-642-79525-1
DOI: 10.1007/978-3-642-79525-1

ISSN 0-387-58848-5 (New York ...)
NE: Reuther, Rudolf [Hrsg.]

Softcover reprint of the hardcover 1st edition 1996

Cover Design: Struve & Partner, Heidelberg

SPIN: 10129709 32/3136 – 5 4 3 2 1 0 – Printed on acid free paper

Preface

It is not long ago that scientists realized, our study and understanding of most environmental problems call for a cross-sectional, more holistic view. In fact, environmental geochemistry became one of the legs to stand on for such a required interdisciplinary approach. Geochemists do not only describe the elemental composition and processes of natural systems, such as soils, ground or surface waters, but they also establish the methodology to quantify material rates and turnover. Today, geochemical expertise has become indispensable when monitoring the fate of noxious chemicals, like metallic pollutants released to the environment. To know how trace metals will behave and react in complex systems under changing conditions, might provide us with a more realistic estimate of what is really acceptable in terms of quality standards. This would ease the formulation of adequate environmental objectives, strategies and criteria to handle emerging pollution situations. Moreover, to take notice of geochemical principles will support our endeavor to improve the way we deal with limited and nonrenewable resources. It is exactly here, i.e. at the interface between natural elemental processes and the way we use them, that geochemical approaches meet the demand of technical attempts to minimize the impact of environmentally relevant activities, like mining, waste handling, or manufacturing.

The consideration to include geochemically derived concepts into the search for technical solutions is not really new, but has a long tradition during the evolution of modern societies. However, it was not until the dispersion of contaminants by human activities reached critical levels that geochemical principles and methods became increasingly involved into existing pollution control and remedial options. Geochemically based monitoring programs proved particularly helpful to elaborate how natural systems may respond upon technical measures, designs or operations, and how this type of knowledge can be applied to predict, avoid or change a conflicting situation. This expertise has become crucial for many private and public sectors, like the production and disposal of residues, the rehabilitation of old waste dumps and industrial sites, or the long-term management of water resources. By utilizing fundamental geochemical processes, such as neutralization, solidification, or concentration, engineers increasingly succeed to manipulate and improve the physico-chemical state and biogeodynamics of contaminated natural or man-made systems.

An overview of this subject is *presented* by R. Schuiling in Chapter 1 on the main geochemical principles, with many examples illustrating the wide range of applications to reduce the mobility, reactivity and toxicity of inorganic waste constituents. It is concluded that geochemically designed engineering is not very energy-consuming, as the inherent energy potential of the processes occuring in waste deposits can be used. Likewise, these techniques are not capital-intensive and do not depend on high-tech, but rather include natural raw materials.

In Part I of this volume, *Criteria and Methods*, the growing consideration of analytical techniques developed to assess the leachability, or short and long-term behavior of metals, e.g., in waste depositories, contaminated soils, or aquatic sediment, is indicated. The need to improve these approaches is discussed by A. Hirner, who presents a new combined elution test for both inorganic and organic forms of metals. R. Reuther claims in Chapter 3 that criteria and standards for metals in water, soil, and sediments must incorporate principles of chemical speciation in order to provide a more realistic estimate for risk assessments. The early identification of critical metal species associated with technical processes or products may help engineers to focus in time on the crucial issues. The use of a rather straightforward speciation method to distinguish between the main forms of mercury in soils from the mining area around Idrija, Slovenia, is demonstrated by H. Biester et al.. The results underline the capacity of these approaches to trace both sources and paths of heavy metals.

In Part II, *Processes in Soils and Lakes*, some practical examples depict how geochemical methods can be used to monitor the release and fate of metals in soils and waters. E. Helios-Rybicka summarizes the main geochemical processes and factors governing the release of metals into rivers, groundwater, and soils from coal and metal mining activities in Poland. The question whether or not metals are leached from soils contaminated by the copper industry of Legnica and Glogow (SW Poland), and what factors may control their mobility is examined by A. Karczewska, who uses a chemical extraction scheme to address this subject. A major issue is the acknowledgement that existing quality criteria and background values for metals are by no means universally valid. Whether a contaminated site should be cleaned-up or left untreated may not only be decided upon in compliance with existing threshold values. This is reflected by J. Matschullat's geochemical inventory of contaminated soils within a lead processing factory. It is advocated that it would be less harmful and less expensive in this case to contain the contaminated soil material within the factory, which would in addition save valuable disposal capacity. The necessity of site-specific decisions versus a global application of standards is stated.

ing lakes in the Lusatian and central German brown coal area. Main geochemical parameters (i.e. the buffering capacity) and reactions (i.e. acidification due to sulfide oxidation) are described in relation to technical measures to restore these lakes (e.g., by liming, river throughflow, biological production). Results from an extensive field survey in 95 Swedish lakes are interpreted by L. Håkanson by means of a new mixed dynamic model. This model is able to describe and predict main biological and geochemical processes and factors (e.g., distribution and bio-uptake) in lakes, which have been treated by different remedial measures (liming, potash and selenium addition, and intensive fishing) to reduce the content of mercury and radiocesium in fish.

Part III of this volume, *Waste Treatment and Landfill Disposal*, is dedicated to problem solutions with regard to the long-term effect potential of metals in waste products and landfills. Two chapters review recent practical developments and implications arising between solid waste treatment, reuse and disposal options, and the use of geochemical concepts. U. Förstner examines critical geochemical processes which describe the mobilization of metals from waste materials. In order to understand and realize landfill stabilization and control, four different geochemical concepts are formulated, i.e., for processes (*mobility of metals*), effects (*storage capacity controlling properties* and *coupled geochemical cycles*), and problem solutions (to achieve *final storage quality*). Available test procedures are discussed for short- and long-term prognosis of the behavior of metals in solid residues. Derived from the main prognostic parameters, i.e., acid- producing potential, storage capacity, and extractability, examples are given for the various techniques for treatment, safeguarding and sanitation of solid wastes and contaminated sites. G. Fleming reviews in Chapter 11 the perspectives of geochemically based engineering techniques to promote and encourage good landfill practices, in light of existing international legislation. In order to be successful, geochemical engineering techniques require a good understanding of the processes and composition – which is by far not constant – within a landfill. The transport mechanisms by which contaminants may move through geochemical barriers are discussed. Examples are given to demonstrate that geochemical expertise can be very helpful to reach a final storage quality as soon as possible, the primary goal of every landfill management.

This volume as a whole shows that close cooperation of earth scientists, biologists, and chemists with engineers may result in solutions which reduce the reactivity of metals in technical products and processes. The transformation of the necessary knowledge into practical environmental measures may also enhance our awareness of the limited capacity of natural systems to counteract or to recover from pollutant stress.

All contributors are thanked for their willing cooperation and support for this book project. The generous and kind engagement of Dr. Wolfgang Engel and Andrea Weber from Springer Verlag is especially acknowledged.

January 1996, Teningen Rudolf Reuther

Table of Contents

INTRODUCTION

PART 1 CRITERIA AND METHODS

PART 2
PROCESSES IN SOILS AND LAKES

List of Contributors

Biester, Dr. Harald
Institute for Environmental Geochemistry
University of Heidelberg
Im Neuenheimer Feld 236
D-69120 Heidelberg
Germany

Fleming, Prof. George
University of Strathclyde
Division of Water Engineering & Environmental Management
Dept. of Civil Engineering
John Anderson Building
107 Rottenrow
Glasgow G4 0NG
United Kingdom

Förstner, Prof. Ulrich
Division of Environmental Engineering
Technical University of Hamburg-Harburg
Eißendorfer Str. 40
D-21073 Hamburg
Germany

Geller, Prof. Walter
UFZ-Centre for Environmental Research Leipzig-Halle
Department of Inland Water Research Magdeburg
Am Biederitzer Busch 12
D-39114 Magdeburg
Germany

Håkanson, Prof. Lars
Institute of Earth Sciences
Uppsala University
Norbyvägen 18B
S-752 36 Uppsala
Sweden

Helios-Rybicka, Prof. Edeltrauda
University of Mining and Metallurgy
Faculty of Geology, Geophysics and Environmental Protection
Al. Mickiewicza 30
PL–30-059 Krakow
Poland

Hess, Dr. Axel
Institute for Environmental Geochemistry
University of Heidelberg
Im Neuenheimer Feld 236
D-69120 Heidelberg
Germany

Hirner, Prof. Alfred V.
Institute of Environmental Analytical Chemistry
Universitätsstrasse 3–5
University of Essen
D-45141 Essen
Germany

Karczewska, Dr. Anna
Institute of Soil Science and Agricultural Environment Protection
Agricultural University of Wrocław
Grunwaldzka 53
PL–50-357 Wrocław
Poland

Matschullat, Dr. Jörg
Institute for Environmental Geochemistry
University of Heidelberg
Im Neuenheimer Feld 236
D-69120 Heidelberg
Germany

Müller, Prof. German
Institute for Environmental Geochemistry
University of Heidelberg
Im Neuenheimer Feld 236
D-69120 Heidelberg
Germany

Reuther, Dr. Rudolf
Environmental Geochemistry and Consulting
Hindenburgstr. 17
D-97320 Albertshofen
Germany

Schuiling, Prof. Roelof "Olaf" D.
Institute of Earth Science
University of Utrecht
P.O. Box 80021
NL-3508 TA Utrecht
The Netherlands

Schultze, Martin
UFZ-Centre for Environmental Research Leipzig-Halle
Department of Inland Water Research Magdeburg
Am Biederitzer Busch 12
D-39114 Magdeburg
Germany

Introduction

1 Geochemical Engineering: Principles and Case Studies

R.D. Schuiling

1.1 Introduction

It is a common misconception that environmental problems have started with the advent of industrialization; situations that we would now describe as environmental problems have existed on earth since its origin. We need only think of the large amounts of SO_2 and other toxic gases that are emitted by certain volcanoes. Similar point emissions in an industrial plant would almost certainly cause the plant to be closed for environmental reasons. It is obvious that the accumulated emissions of all volcanoes during geologic time would have seriously damaged the earth's ecosystems unless the damage were undone by geochemical processes.

Heavy metal "pollutions" have also formed many times in geologic history, however, these metal pollutions are described by economic geologists as ore-bodies. All ore-forming processes have in common the mobilization of the low metal content of a large reservoir by some agent and the redeposition in a small, high-grade ore deposit. Weathering and erosion of such deposits will at some later stage cause such concentrations to be dispersed again.

Rivers transport large amounts of dissolved substances to the sea. The residence times of every element are short with respect to the age of the oceans, yet the oceans are not a saturated solution for most of these elements. This means that they are eliminated from the ocean systems by other mechanisms than supersaturation and precipitation of one of their compounds. Most trace elements are incorporated by isomorphic substitution in the lattice of compounds of major elements, and thereby immobilized. Immobilization mechanisms like these keep the concentrations of many elements in the oceans at very low levels.

Nature thus provides us with many environmentally sound examples of how wastes can be treated. Imitating nature to solve environmental problems, however, has one serious drawback: most geochemical processes proceed very slowly. In order to transform such processes into technologies, the kinetics must be improved by many orders of magnitude. This can often be achieved by increasing the temperature, increasing the surface area of the reactants (e.g. by grinding), and using more concentrated chemicals. In the following, we will give some examples of geochemical engineering, as applied to wastes from industries or mining, or to agricultural wastes.

1.2 Case studies

1.2.1 Breakdown or Neutralization

Starting from our example in the previous section, it is clear that acids in nature are neutralized during chemical weathering. The weathering reactions can be generalized as follows:

acid + water + fresh rock → solid residue (soil) + cations + anion (from acid)

By analogy, many waste industrial acids are neutralized by limestone, but in the case of sulphuric acid, this results in large accumulations of polluted gypsum. Geochemical engineering concepts can circumvent this problem in two different ways:

1. Instead of bringing limestone to the sulphuric acid, the waste sulphuric acid can be injected in limestones underground, where the acid will be neutralized. It has been proved that the heavy metals are immobilized at the reaction front between gypsum and limestone. Moreover, one can take advantage of the fact that the gypsum that forms has twice the volume of the original calcite, which will cause a local elevation of the ground. It can be calculated, and it has been proven experimentally that the reaction and the associated volume increase will take place, regardless of large confining pressures. This swelling effect can be put to good use in the coastal defence of low-lying areas (Schuiling 1989). Possible clogging of the formation close to the bore-hole can be prevented by increasing the injection rate close to the hole; as the kinetics of dissolution of limestone are faster than the precipitation of gypsum, most of the gypsum will precipitate at some distance from the injection, where the radial flow velocity has decreased. Another possibility is to precede the sulphuric acid by hydrochloric acid, which will result in the formation of a $CaCl_2$-brine around the hole. Subsequent injection of sulphuric acid will produce gypsum, and the hydrochloric acid front will be radially displaced to dissolve new limestone.

2. Another geochemical engineering solution is to substitute limestone for neutralization by a different rock-type which will not produce large amounts of solid waste, but will produce some useful by-products instead. It is well known that among the common silicates olivine is the easiest weatherable, and, therefore, experiments have been carried out using ground dunites (olivine-rich rocks) as a neutralizing reactant (Schuiling et al. 1985). At slightly elevated temperatures, around 85°C, the reaction of acid with ground olivine rock proceeds fairly rapidly, up to a pH of about 2 to 2.5. If one starts with a pH of the original waste acid around 0, this means that olivine can neutralize 99% of the acid. The silica that is released from the olivine during the neutralization polymerizes, and separates out as a clean precipitated silica, from which the adhering acid salt solution can be removed by washing. After washing and further neutralization, this precipitated silica with high specific surface area can be used in industry for a number of applications, i.e. fillers in rubber, or as an additive in high-strength concrete. If the acid contains high concentrations of e.g. Ti, Al, Cr and V, like in the waste acids of the TiO_2 -factories, these can be precipitated in a small, but highly concentrated waste stream at pH ~4. Further neutralization of the remaining liquid under admission of air produces fine-grained magnetite. Another useful product, epsomite ($MgSO_4 \cdot 7H_2O$), crystallizes from the residual liquid on cooling. Epsomite can be used to prepare $MgSO_4$ as a commercial product. It can also be used as a starting material for the recycling of sulphuric acid by thermal dissociation of $MgSO_4$ with by-production of magnesia. The fact that all the bulk products are usable makes the olivine process almost into a no-waste environmental technology. In contrast to the normal production of precipitated silica by melting quartz and soda, or by dissolving quartz in autoclaves in alkaline solutions, the olivine reaction is exothermic and can sustain itself once started. Research is now underway to use technical grade acids in the olivine process to industrially produce precipitated silica.

1.2.2 Concentration

Many groundwaters in arid climates contain high levels of fluorine, with serious consequences for public health. People that have to rely on these waters for drinking water may suffer from fluorosis, which affects an estimated 80 million people all over the world. Fluorosis is an illness that first affects the teeth, but in its severe forms causes crippling malformations of the skeleton. Most of the high-fluorine groundwaters are poor in calcium, which is understandable, since fluorite (CaF_2) is fairly insoluble, so calcium and fluorine cannot simultaneously reach high concentrations. If one passes such low-calcium waters before consumption through a filter-bed of gypsum (see Fig. 1.1), most of the fluorine will be eliminated by precipitation of fluorite, according to the reaction:

$$CaSO_4 \cdot 2H_2O + 2F^- \rightarrow CaF_2 + SO_4^{2-} + 2H_2O$$

This is an example of a solution by concentration (de Graaff 1991; Schuiling 1991a). For a description of this and other defluoridation methods, we refer to Heidweiller, 1991. A gypsum filter-bed could also be used as a simple and cheap treatment of fluorine-rich effluents from industry.

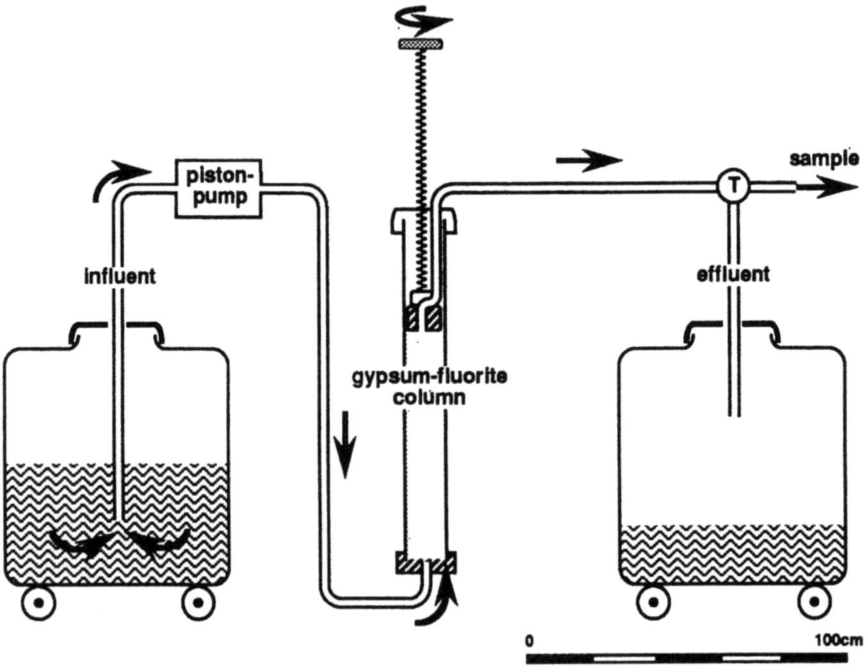

Fig. 1.1. Defluoridation of groundwaters by a gypsum filter-bed

Another example is the following. Many liquid organic waste streams (manure, wastes from milk or potato industries) contain high levels of phosphates. These can be reduced by adding iron or aluminium salts, or by precipitating the phosphate as amorphous calcium-phosphates, but this produces solid phosphate-rich wastes which must be disposed of. It has been observed that anaerobic digestion of accumulations of organic waste in nature sometimes produces struvite, an ammonium-magnesium phosphate. A spectacular example was discovered in the centre of Amsterdam, where accumulations of large struvite crystals were discovered during the excavation of a 13th century site (Kars et al. 1983). Most of the waste streams under consideration have sufficient ammonium and total phosphate, but are lacking in magnesium. Moreover, their pH is in a range where most of the phosphorus is not present as the PO_4^{3-} -ion but as HPO_4^{2-} or H_2PO4^-. The addition of a MgO-slurry overcomes both of these problems, as it raises pH and Mg-concentration at the same time. After the addition of the slurry, the pH slowly starts to rise, and at values between 9 and 9.5, small crystals of struvite are formed, which settle easily and cause no problems in filtering and dewatering. By struvite formation we have succeeded in eliminating most of the phosphates from several liquid organic wastes (Schuiling 1990b), according to the reaction:

$$H_2PO_4^- + Mg^{2+} + NH_4^+ + 6H_2O \rightarrow NH_4MgPO_4 \cdot 6H_2O + 2H^+$$

The resulting struvite is a valuable fertilizer material. If one starts with denitrified manure, the product is a potassium-struvite. This material has been extensively tested in the 1960s as a fertilizer, both in Germany and in the United States, with excellent results. As it had to be synthesized from the pure chemicals, the product was too expensive. In the case of an environmental technology for the purification of waste streams, in which potassium, ammonium and phosphate all have a negative value, the economy of struvite production is considerably improved. Before its use as a fertilizer, it may be advisable to sterilize the product because it may contain germs. In order to sterilize the struvite, one can heat it to 120°C, when it loses five of its six bound water molecules and becomes odorless.

1.2.3 Dilution

Dilution is, of course, so common as an environmental solution to pollution, that it is hardly ever treated as a technology. Every time we wash something, and every time that stack gases are mixed with large volumes of air, we make use of dilution as the means to bring the concentrations of our pollutants down to harmless levels. We will describe an as of yet hypothetical solution to the problem of excess manure production, that is common to regions with a large bio-industry. If the manure production exceeds the receiving capacity of the locally available farmland, the environment is damaged by excess phosphates, causing eutrophication, and by excess ammonia which converts to nitric acid in the air and contributes to acid rain, or oxidizes in the soil and causes high nitrate levels in underlying aquifers. On the other hand, there are many areas in the world which suffer from nutrient deficiencies, including the vast majority of the world's oceans. The surface waters of large parts of the oceans are lacking nutrients to such a degree, that they are described as biological deserts. Oil tankers, returning in ballast to the Middle East, could be used to ship the excess manure. If the manure is spread in the ocean over such biological desert areas, we could increase the biomass in the oceans. If we apply the principle of dilution in this way, the excess manure problem would be solved, and we would fix more CO_2 in the form of oceanic biomass, thereby counteracting the greenhouse effect. Such a proposal will probably meet considerable resistance, as people would consider it equivalent to dumping wastes in the ocean. It is obvious that a large-scale experiment as proposed here, of the order of 5 million tonnes annually should, therefore, be carefully supervised and controlled by marine geochemists, biologists and fishery experts. Even a casual look at the economics of this proposal shows that it will come much cheaper than treating manure in treatment plants on land, by evaporation of their water content, as is presently introduced in the Netherlands.

1.2.4 Isolation

Isolation is the most common environmental option, and it is applied to many mining and metallurgical wastes. Tailing ponds, tailing dams, disposal sites, the backfilling of mined-out stopes, or the disposal in underground cavities are all variations on this theme. To earth scientists, it will be obvious that isolation can never provide a permanent solution to a pollution problem, unless the toxicity of the material to be isolated decreases with time. This is the case with radioactive material or with degradable organic compounds. As isolation will continue to be applied, one can at least slow down the release of toxic substances to the environment by providing the isolated waste deposit with liners that will retard their passage. Highly absorbent clays may serve this purpose, particularly in the presence of carbonates, which will impose a moderately high pH on the solutions with which they come in contact. This will reduce the mobility of many heavy metals considerably. This amounts to saying that the isolation of dangerous pollutants (e.g. nuclear waste) should be accompanied by immobilization measures.

A sensational case of isolation was discovered in the Crimea (Schuiling and van Gaans, accepted). Here a large producer of TiO_2 -pigments and fertilizers disposes of its liquid effluents, amounting to 11 million m^3 annually, by discharging them into an artificial lake of 42 km^2. Thanks to the fact that the climate is dry, discharge of liquids and evaporation will balance for a certain surface area. The effluents, however, are not water but diluted and polluted weak acids. Therefore, the lake becomes more and more acidic during evaporation, and pollutant concentrations increase. The pH of the lake in the course of 28 years operation has decreased to a value of ~0.9. Possible environmental consequences, such as leakage to groundwater, are monitored in approx. 300 monitoring wells. So far no evidence has been found that any acid has leaked into the ground. The explanation of this surprising and comforting result lies in the fact that the lake forms its own protective isolation, by reaction with the underlying marine, carbonate-rich clays. Below a layer of jarosite, that has precipitated on the lake bottom, a very strong thin brown crust was discovered, consisting of amorphous iron-hydroxides and gypsum. After breaking through this impermeable layer, with a thickness of no more than 2–3 cm, the pH of the pore water jumps from a value of 0.9 in the jarosite to 6.2 in the clays. Apparently this reaction layer very efficiently prevents any acid from the lake to escape into the ground. The layer itself shows elevated contents of arsenic, vanadium and chromium, which are effectively immobilized in the iron-hydroxides. Evidently such a *self-healing* isolation is much more effective than any man-made barriers, which are known to fail after a shorter or longer period. If chemical barriers like those described above would fail, these will self-repair by the continued reaction between the two contrasting media. One can envisage isolation policies for toxic wastes, that are based on the formation of such impermeable, isolating layers between the waste and the underlying material, or even between two contrasting waste types.

1.2.5 Immobilization

In environmental technology the term "immobilization" is mostly applied to the physical immobilization of the waste by melting it to a glass, or cementing it in such a way that solutions cannot readily come in contact with the waste. In geochemical engineering, in analogy with nature, it means more often chemical immobilization. This aims at incorporating the pollutants in stable minerals; although this does not diminish their concentrations, it prevents them from entering the biosphere, or allows them to enter the biosphere only at such low concentrations and rates that no harmful effects result. Chemical immobilization may take the form of strong adsorption on minerals like clays or zeolites, or the incorporation in the lattices of stable minerals.

Fly ashes from coal-powered electricity plants are moderately high in various heavy metals, which makes their use in building materials somewhat suspect, as these metals are leachable to various degrees. Such environmental problems can largely be circumvented by pretreating these fly ashes, and the final product may then even be used to solve other environmental problems. Fly ash shows many similarities with volcanic ashes in bulk chemical composition and grain-size distribution. It is known from nature that many volcanic ash beds can be transformed to zeolites by the interaction with hot alkaline waters. Therefore, the fly ash, which consists mostly of reactive glass spherules, has been treated at moderate temperatures with alkaline solutions. Hereby the fly ash is converted to a large degree to zeolites. During this conversion the specific surface increases up to hundredfold and the CEC (cation exchange capacity) by a factor between 100 and 200. This means that the fly ash after treatment is not only capable of retaining its own metal cations, but has a large overcapacity to absorb other pollutants as well. A significant part of the mobile oxy-anions from the fly ash, like molybdates, are dissolved in the alkaline mother liquor. This means that after some cycles this solution must be treated to remove these metals. By the reduction of the heavy metal content, the environmental quality of the zeolitized fly ash is further improved. Figure 1.2 (on p. 10) depicts a fly ash before and after zeolitization.

Fig. 1.2. a Electron-microscopic image of a fresh fly ash

Fig. 1.2. b Electron-microscopic image of a zeolitized fly ash after 24 h in a hot alkaline solution

The SYNROC process (Ringwood 1978) for treating nuclear wastes is another example of immobilization. Certain chemicals are added to the nuclear waste, which is then sintered and produces a mixture of minerals (a SYNthetic ROCk) that are known to be stable under the range of geological conditions that can be expected at the disposal site.

A more mundane example of immobilization is the simultaneous removal of iron and arsenic from groundwater. If reduced groundwater is sprayed over a sand filter, it will oxidize during the intensive contact with air. The ferrous ion will oxidize and precipitate as an amorphous ferric hydroxide; the arsenic, that is present in the arsenite form in the reduced groundwater will oxidize to the arsenate form, and co-precipitate with the iron, as the $Fe^{III}As^{V}O_4$ forms a very stable complex under oxidizing conditions. We dispose of a number of natural examples in the Netherlands in which upwelling groundwaters have precipitated arsenic-rich ferric hydroxides, without causing any environmental harm (Zuurdeeg et al. 1990). Arsenic concentrations in urine of a group of boys living on such arsenic-rich iron-hydroxide soils were even *lower* than those for a control group living on normal soils! This is exactly what one can expect from immobilized situations in contrast to non-immobilized "normal" situations.

The quality of an immobilization technology depends often on the local environment and on maintaining the same local environmental conditions in time. It is well known that polluted harbour sludges (aquatic soils) can contain high levels of heavy metals. Under the reducing conditions in the harbour sludge, these metals are immobilized, probably as very insoluble sulphides. If this material would be dredged and exposed to air, the sulphides would oxidize and become mobilized. A politically controversial, but scientifically sound way to dispose of metal-rich sludges would be to mix them with an excess of sulphide-rich reducing waters, thereby precipitating the metals as insoluble sulphides. Nature's largest body of reducing, sulphide-rich waters is formed by the bottom waters of the Black Sea, which contain very fine metal sulphide particles in suspension. If metal-rich sludges would be pumped into the deep waters of the Black Sea (below 200 m), these would be converted to metal sulphides and slowly settle to the bottom, where they would join a large amount of natural metal sulphides already present in the bottom muds. The reducing water body as well as the low-permeability mud itself will insure that reducing conditions will be maintained over geological periods, forming the modern equivalent of metal-rich black shales. As indicated before, it is unlikely that such a solution will be accepted by policy makers. From the viewpoint of geochemical engineering, however, it is a safe and cheap sanitation technology for metal-rich sludges. In addition, it could bring some financial relief to the countries adjoining the Black Sea, that should be paid for the granting of dumping rights.

1.3 General Considerations

Geochemical engineering solutions have in common that they are usually not very energy-consuming, are not dependent on high-tech, and make use of natural raw materials. These solutions are on the whole not very capital-intensive, but may be labour-intensive. This makes such approaches particularly attractive in economically depressed countries, or more generally in developing countries (Schuiling 1991b). Geochemical engineering aims at minimizing waste production, and tries to transform wastes into useful by-products. A major advantage of geochemical engineering is the fact that closely related processes have been active in nature for many millions of years, which gives us some guarantee that the process will not have unexpected negative effects.

References

Graaff JWM de (1991) Decentralized defluoridation; a gypsum-fluorite filter for the removal of fluoride from drinking water (in Dutch). Report Utrecht University/IWACO BV, p 54

Heidweiller VML (1990) Fluoride removal methods. p.51-85 In: Frencken JE (ed) Endemic fluorosis in developing countries

Kars H, Wevers JMAR, Schuiling RD (1983) Struvite, a mineralogical curiosity from an archeaological site. Berichten Rijksdienst Oudheidkundig Bodemonderzoek, 30:423-426

Ringwood AE (1978) Safe disposal of high level nuclear reactor wastes: a new strategy. Austr Natl Univ Press, Canberra, 64 pp

Schuiling RD, v Herk J, Pietersen HS (1985) A method for neutralizing waste sulfuric acid by adding a silicate. Eur Patent Appl no 8590343.5

Schuiling RD (1993) Procedure for locally raising the ground artificially. European Patent PCT/NL89/00014

Schuiling RD (1990a) Geochemical engineering: some thoughts on a new research field. Appl Geoch 5:251-262

Schuiling RD (1990b) Dephosphatization of aerobically treated calf's manure by the KMP-process (in Dutch). Report Ministry of Agriculture, Netherlands, 37 pp

Schuiling RD (1991a) Decentralized in situ defluoridation (in Dutch). Final report Ministry of Development Cooperation, Netherlands, 6 pp

Schuiling RD (1991b) Geochemical engineering and its potential in developing countries. Proc Int Symp Appl Geoch, Hyderabad, India

Schuiling RD, van Gaans PFM (accepted) Self-sealing of the waste sulfuric acid lake of the TiO_2-plant at Armyansk, Crimea, Ukraine. Appl Geochem

Zuurdeeg BW, Enk R van, Vriend SP (1990) Origin of the elevated arsenic concentrations in soils of the Huet, Doetinchem. A risk analysis (in Dutch). Report GEOCHEM-Research, 68 pp

Part 1

Criteria and Methods

2 Testing Metal Mobility in Soils by Elution Tests

A. V. Hirner

2.1 Introduction

Classification of pedosphere pollution may present severe problems to professional environmentalists, e.g. in practical cases of assessing environmental impact measures associated with waste disposal sites. To enable prompt decisions or remediation actions, usually governmental institutions give strict guidelines for soil pollution assessment.

In Germany, at first instance soil metal pollution grade is based on the total metal content as received by acid dissolution of the sample (named the S7 test; Deutscher Normenausschuß 1984a). To establish mobility and/or bioavailability of inorganic pollutants, there is no doubt that S7 concentrations are too high. Thus, in a second test, the sample is eluted by distilled water for 24 h (S4 test; Deutscher Normenausschuß 1984b). Unfortunately, with this S4 test, the concentrations will clearly be too low, as in the past the samples had usually experienced more aggressive waters for longer time periods.

Metals associated with hydrophobic organics cannot be extracted by aqueous solutions. Therefore, elution tests have to be improved considerably. Furthermore, the presence of humic acids should be accounted for and it would be desirable to apply just one elution test for inorganic as well as organic pollutants simultaneously. Some steps towards these aims (pH_{stat} and two phase elution test) will be discussed in this paper.

2.2 Methods

2.2.1 Sequential Extractions

Single (elution tests) and sequential chemical extractions are capable of differentiating between different metal mobilities (Quevauviller et al. 1993) and thus, have already proved their potential in the field of environmental analytical chemistry; the most used pertinent sequential extraction schemes are discussed in literature (Hirner 1992). From this study (Hirner 1992) an extraction sequence can be derived for low-carbonate soils, avoiding strong interferences between inorganic and organic fractions combined with a relatively high discrimination potential as received by nine fractions (see Table 2.1). It has to be noted, however, that only samples with a maximum carbonate content of 5 wt.-% can be separated on basis of this sequence. Following the first extraction step, together with NH_4OAc the equivalent amount of HCl for carbonate dissolution must be added.

Depending upon the polarity of organic soil constituents and/or working place safety arguments, benzene/methanol may be substituted by other solvents, e.g. toluene/dichloro-methane, acetone, ethanol, and others. Future trends point to the application of supercritical extraction fluids (SFE/SFC), thus avoiding hazardous chemicals.

In order to harmonize extraction procedures, relevant research (Ure et al. 1993) has been performed under the auspices of the BCR (Community Bureau of Reference, Brussels). Unfortunately, a slightly modified version (Legret 1993) of

Table 2.1. Sequential extraction procedure for low-carbonate soils (Hirner 1992; Zeien and Brümmer 1989)

Fraction		Extractant
	Exchangeables	
1.	Unspecific adsorbed	NH_4NO_3
2.	Specific adsorbed	NH_4OAc (+ HCl*)
	Soluble organics	
3.	Solvent extractables	Benzene*/methanol*
4.	Humic and fulvic acids	Benzene*/methanol*/KOH
	Oxides	
5.	Mn-oxides	$NH_2OH \cdot HCl + NH_4OAc$
6.	Amorphous Fe-oxides	$(NH_4)_2C_2O_4$
7.	Crystalline Fe-oxides	Ascorbic acid/oxalate buffer
8.	***Silicates***	HF
9.	***Insoluble organics***	$HNO_3/HClO_4$

* see text

the Tessier extraction scheme (Tessier et al. 1979) that was criticized in the pertinent literature (Kheboian and Bauer 1987; Nirel and Morel 1990; Accomasso et al. 1993) was applied. Thus, in the course of reference experiments upon single mineral phases, the BCR procedure showed poor phase selectivity (Whalley and Grant, 1994).

2.2.2 pH_{stat} Elution Test

In extension of the above mentioned S4 elution test, other tests based on acid-aqueous solutions (often together with complexing agents, e.g. (Crosland et al. 1993) are in practical use in The Netherlands, Switzerland, and the United States.; a summary is given on p. 387 in Ulrich Förstner's book (1993).

A special concept to test the long-term mobility of metal pollutants in waste materials was developed by Obermann and Cremer (1991; see Fig. 2.1): the initial behavior of freshly deposited wastes is tested by S4 elutions. Extrapolations to long-term scales is made on the basis of the results of elutions carried out at a constant pH 11 and constant pH 4, (pH_{stat} test). These conditions mimic those

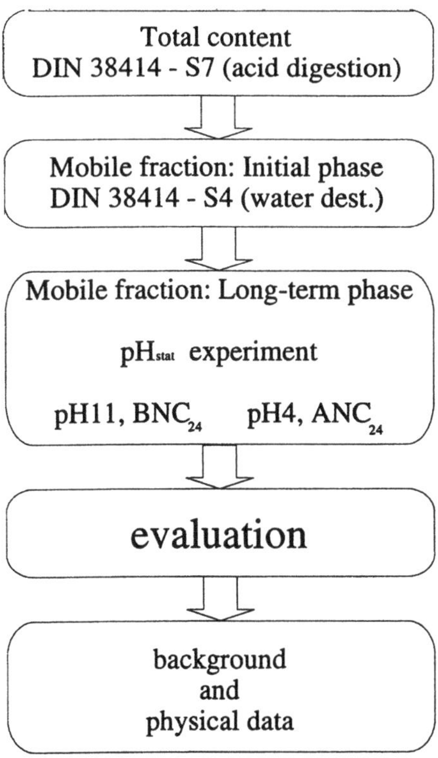

ANC_{24} : acid neutralisation capacity after 24h elution
BNC_{24} : base neutralisation capacity after 24h elution

Fig. 2.1. Elution test to evaluate metal pollutant mobility in waste (Obermann and Cremer 1991)

of strongly basic and acidic waters in waste deposits, and thus, are resembling a "worst case" scenario. Eventually, the probability for this "worst case" is estimated on basis of the measurement results of the base/acid neutralisation capacities.

Automated apparatus are available commercially, where several samples are processed simultaneously by computer controlled titration stands (Schriever 1994). To provide enough turbulence during elution, shaking devices as well as stirring techniques are used.

Furthermore, if suitable criteria are applied to metal concentrations in the pH_{stat} eluate, the final storage quality of organic-free waste material can be evaluated (Hirner and Förstner 1993).

2.2.3 Combined Elution Test

To investigate soils contaminated with metals as well as with organic substances, and to take into account various inorganic/organic interactions within natural samples, metals associated with organic compounds must also be eluted. As des-

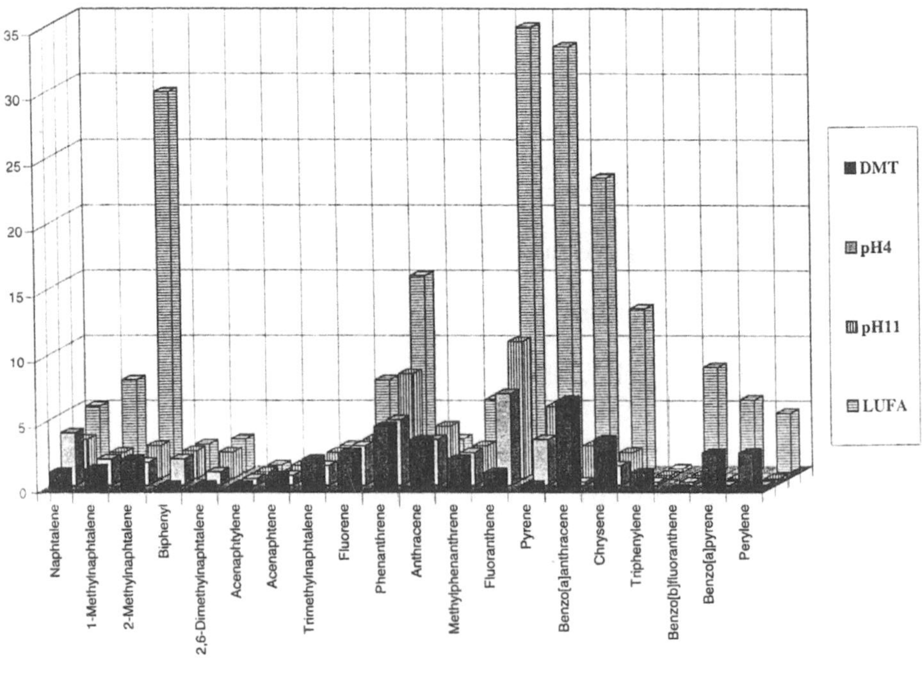

Fig. 2.2. PAHs in a contaminated soil sample as received by different extraction and elution methods

cribed in literature (e.g. Nishimura and Baker 1987), neutral, basic and acidic organic ligands are extractable by organic solvents mixed with basic/acidic aqueous solutions. These mixtures may consist of immiscible (e.g. toluene/water) or miscible liquids (e.g. acetone/water).

Following extraction experiments using toluene, acetone, ethanol and THF, in the Institute of Environmental Analytical Chemistry, University of Essen, a 3:7 (vol.) mixture of ethanol:water (e/w) was applied; basic (pH 11) and acidic conditions (pH 4) were established by addition of NaOH and HNO_3, respectively. Toluene was used as a keeper and organic pollutants were measured by GC-MS in the selected ion monitoring (SIM) mode.

In respect to the concentrations of polycyclic aromatic hydrocarbons (PAHs) in soil, it can be seen in Fig. 2.2, that these elutions show extraction yields of usual extraction procedures in most cases. Concerning the latter, the two methods selected are indicated as "DMT" (Deutsche Montan Technologie) and "LUFA" (Landesuntersuchungsanstalt für Forstwirtschaft und Ackerbau) methods in Fig.2.2. Both these methods are based on acetone extractions and are frequently used in Northrhine Westfalia, Germany.

2.3 Discussion

2.3.1 Problems Associated with Elution Tests

Heterogeneous Sample Composition and Structural Effects. Complex natural samples may be characterized by complicated intercalation of inorganic and organic layers, leading to included phases which cannot be reached by the eluent (Fig. 2.3). In principle, this problem in respect to sequential extractions can be overcome by repetitions of individual extraction steps. The use of elutions-only combined tests may help to solve this problem at least partially.

Dissolution Kinetics. According to Fig. 2.3, under usual experimental conditions (batch experiment), the determination of the eluate concentration is possible in case A only. In case B, long-term flow extractions are necessary (Schoer and Förstner 1987).

Readsorption Effects. Solid surfaces created by the elution procedure can readsorb species from solution. This effect can be minimized, if the elution is not performed in batch experiments, but in flow-through systems (Schoer and Förstner 1987).

Anaerobic Environments. When handling anaerobic samples, extreme care is necessary to avoid contact with oxygen during sampling, transport, storage, and elution (Wallmann et al. 1993). Another method is to simulate the transition from oxic to anoxic environments by elution tests. Although the observed effects may be significant (e.g. release of As and Fe by reduction of Fe(III) to Fe(II); Cu and Cd fixation by sulfide formation), the time necessary for redox experiments is in the order of weeks, as compared to hours typical for pH experiments.

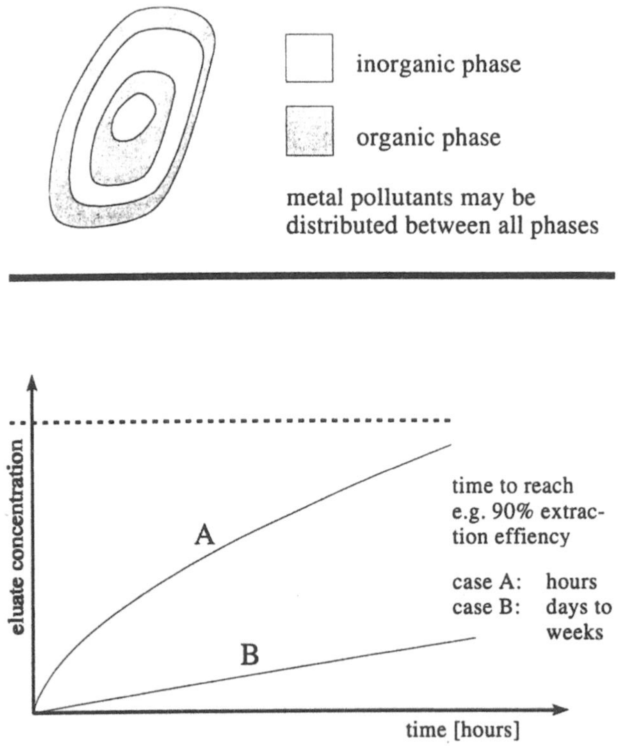

Fig. 2.3. Problems associated with elution of complex natural samples.

2.3.2 Rationale for Comparing Elution Tests

Every elution test is tailored for a specific environmental scenario, as is the pH 4 test for waste deposition (Obermann and Cremer 1991; Hirner and Förstner 1993). The latter is not representing the usual situation in soil chemistry at the earth's surface; the ideal, environmentally relevant elution test is not available yet. This is schematically depicted in Fig. 2.4 together with the unrealistic results of the S4 and S7 elution tests as emphasized in the introduction. The pH 4 test will definitely be rated within the desired target range, as will be the previously described combined elution test. The suitability of the latter test extends such scenarios from waste deposits to oil spills and accidents.

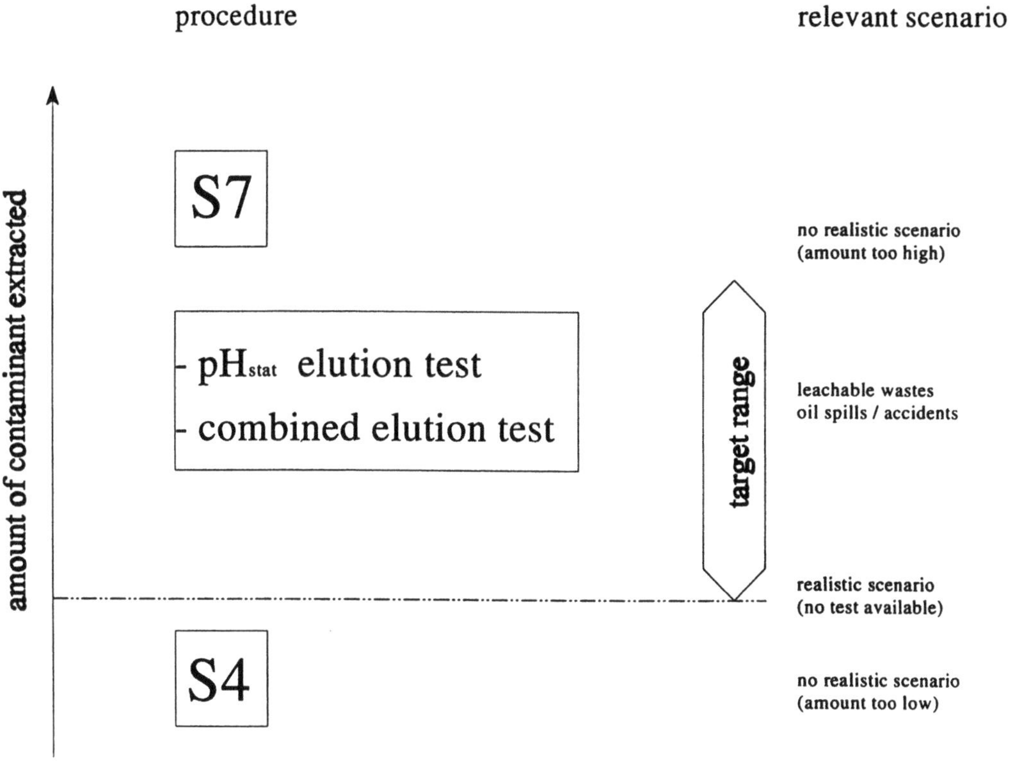

Fig. 2.4. Comparison of elution tests

2.4 Outlook

It is necessary to further test the proposed combined elution procedure with various sample matrices in contaminated soils and waste materials. This is to be done for the elution of metals, as well as the one of important classes of organic pollutants such as CHCs, PAHs, phenols or PCBs. Relevant data will soon be published elsewhere.

The next logical step will be to incorporate natural surfactants into the eluent to simulate more realistically the transport of hydrophobic pollutants in soil solution by micelles. Pertinent experiments are presently performed at the Institute of

Environmental Analytical Chemistry, University of Essen.

Last, but not least, the need for international harmonization (elution procedures, reference samples) has to be addressed (Griepink 1993).

References

Accomasso GM, Zelano V, Daniele PG, Gastaldi D, Ginepro M, Ostacoli G (1993) A study on the reproducibility of Tessier's extractions in a fluvial sediment and a comparison between different dissolution procedures in a reference material. Spectrochim Acta, 49A:1205–1212

Crosland AR, McGrath SP, Lane PW (1993) An interlaboratory comparison of a standardized EDTA extraction procedure for the analysis of available trace elements in two quality control soils. Int J Environ Anal Chem, 51:153–160

Deutscher Normenausschuß (1984a) DIN38414, part 7: Bestimmung des säurelöslichen. Anteils (S7), Berlin

Deutscher Normenausschuß (1984b) DIN38414, part 4: Bestimmung der Eluierbarkeit mit Wasser (S4), Berlin

Förstner U (1993) Umweltschutztechnik. Springer, Berlin Heidelberg New York, 572 pp

Griepink B (1993) Some considerations with regard to the quality of results of analysis of trace element extractable contents in soil and sediment. Int J Environ Anal Chem, 51:123–128

Hirner AV (1992) Trace element speciation in soils and sediments using sequential chemical extractions methods. Int J Environ Anal Chem, 46:77–85

Hirner AV, Förstner U (1993) Elutionstests zur Bestimmung der Schadstoffmobilität. Altlasten, 3:29–30

Kheboian C, Bauer CF (1987) Accuracy of selective extraction procedures for metal speciation in model aquatic sediments. Anal Chem, 59:1417–1423

Legret M (1993) Speciation of heavy metals in sewage sludge and sludge-amended soil. Int J Environ Anal Chem, 51:161–165

Nirel PM, Morel FM (1990) Technical note. Pitfalls of sequential extractions. Wat Res, 24:1055–1056

Nishimura M, Baker EW (1987) Compositional similarities of non-solvent extractable fatty acids from recent marine sediments deposited in different environments. Geochim Cosmochim Acta, 51:1365–1378

Obermann P, Cremer S (1991) Materialien zur Ermittlung und Sanierung von Altlasten, Bd 6; Landesamt für Wasser und Abfall NRW. Düsseldorf, 127 pp

Quevauviller Ph, Ure AM, Muntau H, Griepink B (1993) Single and sequential extraction in sediments and soils. Int J Environ Anal Chem, 51:129–134

Schoer J, Förstner U (1987) Abschätzung der Langzeitbelastung von Grundwasser durch die Ablagerung metallhaltiger Feststoffe. Vom Wasser, 69:23–32

Schriever M (1994) Standort Chemie, 1, 9

Tessier A, Campbell PGC, Bisson M (1979) Sequential extraction procedure for the speciation of particulate trace metals. Anal Chem, 51:844–851

Ure AM, Quevauviller Ph, Muntau H, Griepink B (1993) Speciation of heavy metals in soils and sediments. An account of the improvement and harmonization of extraction techniques undertaken under the auspices of the BCR of the Commision of the European Communities. Int J Environ Anal Chem, 51:135–151

Wallmann K, Kersten M, Gruber J, Förstner U (1993) Artifacts in the determination of trace metal binding forms in anoxic sediments by sequential extraction. Int J Environ Anal Chem, 51:187–200

Whalley C, Grant A (1994) Assessment of the phase selectivity of the European Community Bureau of Reference (BCR) sequential extraction procedure for metals in sediment. Anal Chim Acta, 291:287–295

Zeien H, Brümmer GW (1989) Chemische Extraktionen zur Bestimmung von Schwermetallbindungsformen in Böden. Mitt Dt Bodenkundl Ges, 59:505–510

3 Geochemical Speciation: Does it Help to Assess and Engineer the Impact of Metals?

R. Reuther

3.1 Definitions and Introduction

According to the definition we will use here, a *chemical species* represent 'expressis verbis' the molecular arrangement of a particular metal or metal compound. This may include free aquated ions, distinct chemical compounds (such as dissolved inorganic salts and acids, or solid oxides, sulfides, and carbonates), as well as thermodynamically unstable aggregates (like dissolved and solid organic complexes, colloids or coprecipitates) either naturally available or as a result of man's activity.

The term *speciation* will be defined here both in an operational and in a more descriptive way (Bernhard et al. 1986). *Operational* means the use of appropriate analytical separation and detection methods (e.g. filtration, chemical leaching and extraction, chromatographic fractionation, atomic absorption) designed to simulate conditions, which may favor the "extractability" of a particular metal form, association, or binding substrate.

In a more *descriptive* sense, speciation relates to the characterization of geochemical transport mechanisms, like advection, diffusion, and dispersion leading to the accumulation or deprivation of metals. It also describes transformation phenomena, like adsorption–desorption, precipitation–dissolution, complexation, and alkylation, by which operationally defined metal species may mobilize or bind to various environmental media (Reuther 1987).

In this context, the term *engineering* will refer to technical means developed to manipulate and steer the geochemical status of metals in water, soil and wastes, by either intensifying or weakening those factors which maintain a certain species composition for a given site and time.

An *Environmental Impact Study* (EIS) as regarded today is a regulative measure to protect man and the environment from possible adverse effects of relevant public and private activities. In that sense, it is used as a preventive instrument to identify, describe and evaluate in advance possibly harmful effects which may arise from significant projects, plans, programs, facilities, technologies or products. The results may provide basic data to decide on or change project design in time, before irreversible damage to ecosystems may occur.

With regard to chemical pollutants, the objective of an EIS seems two-fold, i.e. (1) to predict unwanted, possibly adverse consequences of associated chemicals before a project starts (*risk analysis*) and (2) to safeguard project implementation, development, and termination, according to accepted emission and quality criteria (*compliance monitoring*, OECD 1979).

In the following, it is tried to demonstrate the superiority of chemical speciation data over bulk concentrations, in particular when used as assessing criteria to evaluate and predict the long-term effect potential and ecological relevance of metal pollutants.

3.2 Impact Studies of Project-Associated Metal Pollutants

In contrast to single environmental laws, the existing EIS regulation in many countries emphasizes the priority to estimate the whole effect potential associated with a planned project. An early understanding of the *effect-related characteristics* (i.e. solubility, availability, toxicity) of a metal pollutant released by a certain project will help to avoid or mitigate possible adverse impacts. In practice, however, the installation of appropriate guidelines, as claimed by legislation, is still poorly defined and not based on criteria which account for the complex interactions and fluctuations occurring in natural systems.

As an example, § 6 of the German Environmental Impact Assessment Act (EIA) requires from the project owner a priori the submission of all relevant planning data to the licensing authority, which include (i) the type and amount of applied, produced, and released substances as well as (ii) the size and nature of possibly associated effects. Furthermore, a comprehensive inventory of the existing environmental situation at the planned project site is requested, with a description of major elemental fluxes, chemical and biological key parameters, and natural variations. From this the preload, vulnerability, and compatibility of the site with regard to the proposed project has to be estimated (Gesetz zur Umweltverträglichkeitsprüfung 1990).

With regard to the fate of chemical contaminants like heavy metals, one major shortcoming of most EIS systems is their obvious lack of a clear-cut scientific criterion which reflects the *mobility and reactivity* of a metal subjected to particular site conditions.

Compliance with emission and effluent standards is required according to what is technically feasible (e.g. by best available technology), economically reasonable, and ecologically sound regarding to the type and size of a given project and the site characteristics. The formulation of air and water emissions and of established quality criteria for receiving systems (air, water, soil, sediment, waste depository) is directly related to total pollutant concentrations. Consequently, both the requested input–output analysis and the site evaluation refer solely to total concentrations or to the total amount of pollutants released per time interval or production volume. Most of these standard values are simply derived from short-term laboratory tox-tests using bulk elemental concentrations and single organism species.

3.3 Total Concentration Versus Species Concentration

Today it is common scientific understanding, that the environmental behavior of chemicals, like heavy metals, i.e. their reactivity, mobility, and availability in ecosystems, as well as possible effects (in terms of biological harm or benefit), strongly depend on their *particular form* (Förstner and Wittmann 1979). It is also realized that the toxicity of pollutants in natural systems may be higher (or lower) than what is predicted by quality criteria. Emission and exposure standards, still based on total concentrations, do not account for the significant differences existing between the various single forms of an element.

The primary goal of an impact study is to provide basic knowledge about how to prevent, or reduce and control any possible harm to biological systems, e.g. by implementing appropriate, project-related countermeasures (such as emission control, integrated technologies, closed cycles, or compensating measures). With regard to chemical contaminants this ultimate goal may not be attainable unless the ecological significance of speciation effects is adequately addressed. Official threshold values used to estimate the exposure to toxic chemicals do not take advantage of the information which is provided by speciation data or models, particularly with regard to long-term prognosis, secondary effects, or site-specific reactions.

To ensure a successful technical and economic project performance, the actual risk of potentially hazardous components, like heavy metals, must be assessed and minimized. For this, chemical speciation methods can offer a more realistic measure to identify particularly *highly mobile* or *biologically reactive* forms before project start-up.

Ecologically unacceptable forms of metals may include the following criteria (see Positionspapier Gesellschaft Deutscher Chemiker 1993):

- High geochemical mobility
- Easy biological uptake
- High chemical stability
- Long ecological half–life

In contrast to the partitioning of organic micropollutants, like halogenated and polyaromatic hydrocarbons, a wide range of speciation procedures now exists to determine critical forms of metals in various environmental matrices, including dust, water, soil, sediment, and biological tissue (Fig. 3.1; Allen 1993). Yet, chemical speciation procedures are far from being standardized due to their lack of selectivity and reproducibility (Reuther 1996). In addition, monitoring and licensing authorities consider them as too sophisticated, time-consuming, unpractical, nonrepresentative, and too expensive for the daily routine work.

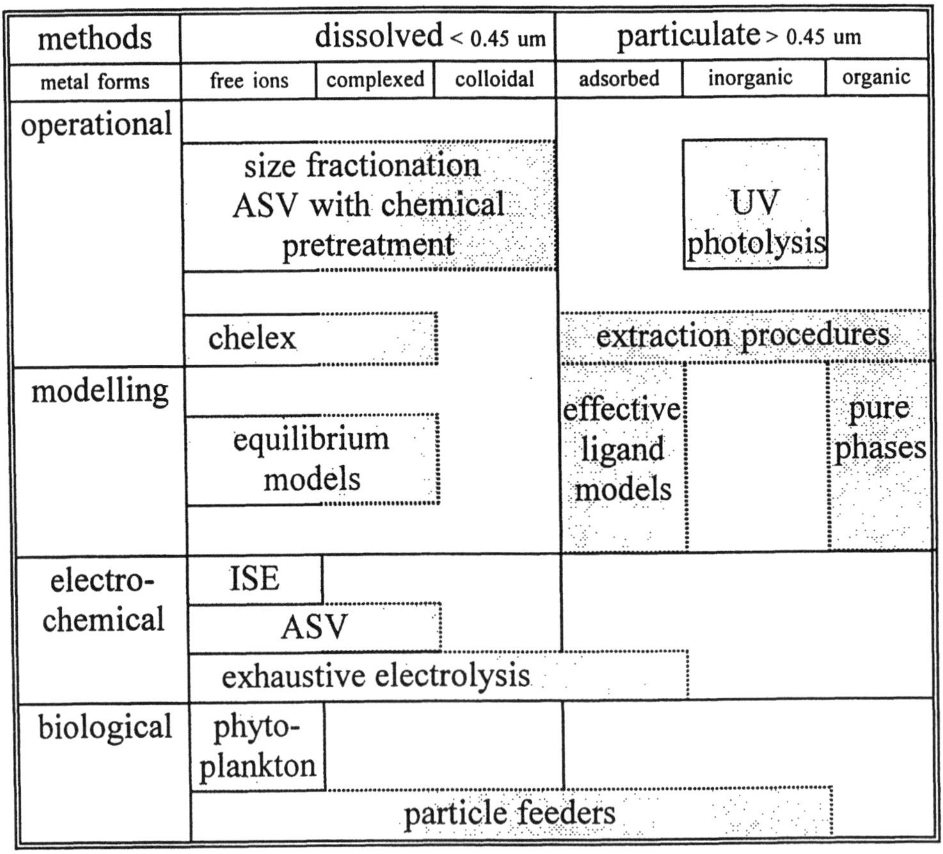

Fig. 3.1. General classification scheme for speciation methods (redrawn after Turner and Whitfield 1980)

3.4 Assessment Approach

During the scoping step of an EIS, a *first ranking* of those components (e.g. metals and organics) is recommended, which are classified as potentially hazardous either due to the amount produced, known physicochemical and toxicological characteristics, or in relation to their application and release pattern, their environmental occurrence, or due to a particular site susceptibility.

Regarding metals (or organic micropollutants), this can be realized by examining and comparing calculated emission and immisssion data with given *limit values* or with officially accepted *guidelines* elaborated for *priority pollutants* in reference lists, like the UN IRPTC, the USEPA National Priorities List, or the German catalogue of water-hazardous substances (see also Voigt and Rohleder 1986; van der Gaag et al. 1991).

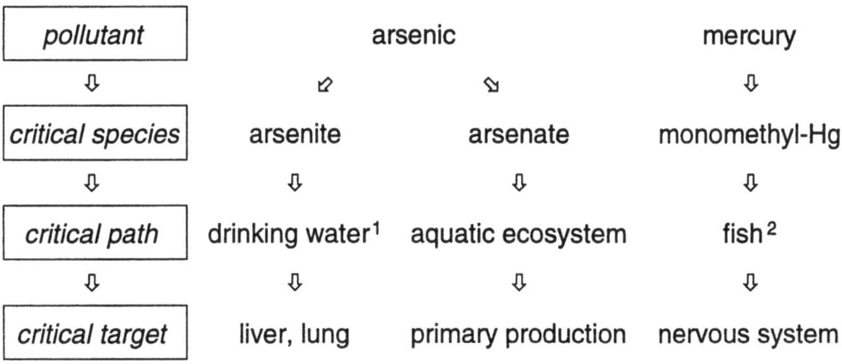

Fig. 3.2. Critical pathway analysis (example) to estimate the impact of arsenic (50 ug/l)[1] and mercury (500 ug/kg[1, 2]) WHO recommended values

When the environmentally most critical components within a complete project cycle are identified (e.g. by their property or total amount), the pathway(s) should be analyzed by which these substances may reach critical concentrations in both man and associated foodwebs. This type of risk estimation, known as *critical pathway analysis*, is primarily based on the evaluation of existing speciation data, as exemplified in Fig. 3.2 for arsenic and mercury (Preston and Portmann 1981). The differentiation of the most toxic forms of both metals unveils rather early the really critical forms and pathways which these metals may take to reach sensitive targets. To focus on total concentrations alone would fail to understand and predict their true nature or to find out the really relevant routes of exposure.

To get an early estimate of the potential (*risk*) or direct threat (*hazard*) by contaminants, like heavy metals, environmental assessments should check at first, if known hazardous elements associated with the project may occur in a critical chemical form (cf Fig. 3.2). Such an examination may be complemented by the use of *bioassays and biomarkers* to assess relevant toxicological endpoints such as organism growth, mobility, survival, or enzyme activities, and to validate the accuracy and biological relevance of the results obtained by chemical speciation.

In the relevant literature, a close relationship between the environmental behavior and the chemical form of heavy metals is documented for *aquatic systems* (Förstner and Wittman 1979). As an example in Fig. 3.3 main chemical forms (weight-%) of nickel (Ni) are depicted against total concentrations (in ppm) in a vertical sediment profile taken from two acidified lakes near the Sudbury mining and smelting area, in Ontario, Canada. Both enrichment and depletion of Ni in the upper sediment layer of Lake Ramsey (located close to the mining area) are caused by the occurrence of mainly ion-exchangeable forms, which decrease together with residual forms as a result of building the "superstack" in 1972. In the acidic Lake Wavy (2.5 km away from Sudbury) reducible Ni species (Ni bound to Fe and Mn oxides) determine the overall behavior of this metal in the sediment. Here also total concentrations decrease in the more recent part of the sediment, although the input of particularly insoluble Ni forms by long-range atmospheric transport from the smelter continues. This is indicated by the increase of the residual fraction in the uppermost layer.

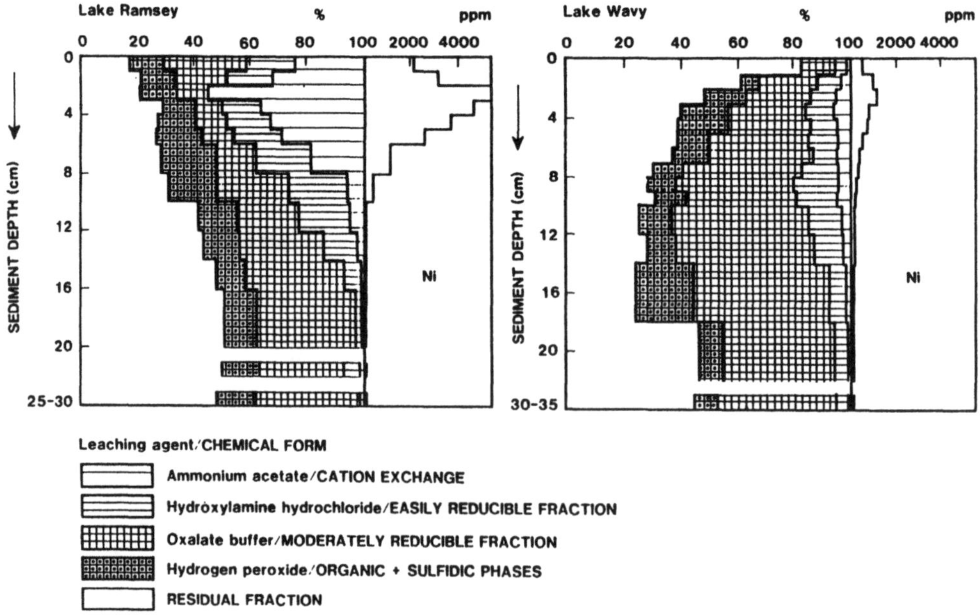

Fig. 3.3. Vertical distribution of chemical forms of nickel (Ni) in lake sediments from the Sudbury mining area, Canada (Reuther 1983)

3.5 Conclusions and Recommendations

The public acceptance of relevant industrial activities, like resource extraction and processing, product manufacturing, refining and use, energy generation and consumption as well as waste treatment and disposal practices (see this volume), is much related to the existing preload, sensitivity, or tolerance of a prospective site or area.

An early understanding of the mobility and long-term behavior of contaminants which are classified to be critical in receiving systems or for a certain type of project, may facilitate and optimize project planning and implementation.

In this context, to focus on critical metal species concentrations, i.e. their temporal and spatial variation, may help to predict in due time disturbances of typical element cycles, e.g. by *changing speciation pattern*, which may arise by a project (see Fig.3.3).

Although still in its beginning, simple leaching tests are now employed by agencies for routine use to monitor "*mobile*" or *acid-soluble* parts of metals, e.g. in solid residues, in addition to total concentrations (see Hirner, Chap. 3 and Förstner, Chap.11). In a first attempt, the chemical extraction of solid waste material with distilled water or weak acids (i.e. 0.1 HCl) has become a standard procedure when assessing the initial leachability of metals in hazardous waste materials (cf German DIN 38414-S4 or US–EPA TCLP-leaching test; see also Förstner 1993).

In the long run it is hoped that the early identification of the critical form of metals associated with a certain type of project or industry may help planers and engineers to find appropriate technical measures to manipulate the behavior of metals in a way, which guarantees an occupationally and environmentally safe project development. Besides, it may save considerable costs by focusing in time on the real crucial issue.

Summarized, it may be concluded that there is a future need:

1. To standardize and improve the analytical quality of existing speciation methods

2. To incorporate principles of metal speciation into established emission standards and quality criteria particularly for high-priority elements, in order to provide more realistic numeric values

3. To include sector or site-specific geochemical speciation data into the design of environmental engineering measures (e.g. liming, drainage, solidification, landfill waste disposal, effluent treatment)

4. To use geochemical speciation procedures in combination with bioassays and epidemiological data

References

Allen HE (1993) The significance of trace metal speciation for water, sediment and soil quality criteria and standards. The Science of the Total Environment, Supplement 1993, Elsevier, 23–45

Anonymous (1986) National priorities list fact book. HW 7.3, US Environmental Protection Agency, Washington DC, 94 pp

Anonymous (1987) International register of potentially toxic chemicals (IRPTC), Legal File 1986, Vol. I and II, United Nations Environment Program

Bernhard M, Brinckman FE, Sadler PS (eds) (1986) The importance of chemical speciation in environmental processes. Dahlem Conferences, September 2–7, 1984, Springer, Berlin Heidelberg New York

Förstner U (1993) Schlacke und Endlagerqualität, Konzepte, Bewertungskriterien, Perspektiven, manuscript, 35 pp

Förstner U, Wittman G (1979) Metal pollution in the aquatic environment, Springer, Berlin Heidelberg New York

Förstner U (1996) Waste treatment: geochemical engineering approach (this volume)

Gesellschaft Deutscher Chemiker (1993) Grundsätze der ökotoxikologischen Bewertung von Chemikalien und Sachstand der heutigen Praxis. Positionspapier Fachgruppe Umweltchemie und ökotoxikologie, Arbeitskreis Chemikalienbewertung

Gesetz zur Umweltverträglichkeit (Environmental Impact Assessment Act) (1990) Bundesgesetzblatt

Hirner AV (1994) Testing metal mobility in soils by elution tests (this volume)

OECD 1979. Environmental Impact Assessment, Paris, 71 pp

Preston A, Portmann JE (1981) Critical path analysis applied to the control of mercury inputs to United Kingdom coastal waters. Environmental Pollution (Series) 2:451–464

Reuther R (1983) Anreicherung und Bindung von Schwermetallen in See–Sedimenten unter dem Einfluß saurer Niederschläge. PhD Thesis, University of Heidelberg, 127 pp

Reuther R (1987) The metal conference in Athens, 1985: a growing interest in metal speciation; a review. In: Landner, L (ed), Speciation of metals in water, sediment and soil systems. Lecture Notes in Earth Sciences, Springer, Berlin Heidelberg New York

Reuther R (1996) Trace metal speciation in aquatic sediments: methods, benefits and limitations. In: Mudroch A, Azcue JM, Mudroch P (eds) Manual of physico-chemical analysis and bioassessment of aquatic sediments, Lewis (in print)

Salomons W, Stigliani WM (1995) Biogeodynamics of pollutants in soils and sediments. Environmental Science, Springer, Berlin Heidelberg New York

Turner DR, Whitfield M (1980) Chemical definition of the biological available fraction of trace metals in natural waters. Thalassia Jugosl. 16:231–241

Van der Gaag MA, Stortelder PBM, van der Kooij LA, Bruggeman WA (1991) Setting environmental quality criteria for water and sediment in The Netherlands: a pragmatic ecotoxicological approach. European Water Pollution Control 1 (3):13–20

Voigt K, Rohleder H (1986) Datenquellen für Umweltchemikalien: Monographien, Nachschlagewerke, Datenbanken, Ecomed Verlagsgesellschaft, 122 pp

4 Investigations on Different Mercury-Phases in Soils of a Mercury-Mining Area by a Pyrolysis Technique

H. Biester, A. Hess and G. Müller

4.1 Introduction

Mercury (Hg) contamination has attracted more attention than many other trace elements in the past few years. Local contamination due to mining activities and to industrial use of Hg, as well as increased deposition into the soils of regions far from industrial emission sources, has been detected (Lindquist et al. 1991; Glass et al. 1991; Iverfeld 1991). The emission of Hg is mainly caused by the combustion of fossil fuels, by waste incineration, by chlor-alkali plants, and by the processing of other metals (Pacyna and Münch 1991). The hazard potential of soil contaminations depends mainly on the type of bond, which controls toxicity and mobility. Due to redox- and pH-conditions of the soil and to the adsorption capacity of the different soil components, Hg can be associated with clay minerals, iron oxides, or organic material (Anderson 1979). The strong affinity of Hg to organic materials, especially to humic acids, has been shown by many authors (e.g. Strohal and Huljev 1971; Kerndorff and Schnitzer 1980; Johannsson et al. 1991; Mierle and Ingram 1991). Natural Hg anomalies are usually related to ore deposits where Hg occurs predominantly as cinnabar. Cinnabar shows very low solubility and strong resistance against weathering; thus, the bioavailibility of this form is low.

Mercury-bonds in soils are often determined by various kinds of sequential leaching procedures (e. g. Peng and Wang 1985; Di Giulio and Ryan 1987; Revis et al. 1989). Hence, there is a problem with insufficient selectivity of the extractants (Dües 1987; Papp et al. 1991). Moreover, the application of different leaching methods on the same sample may lead to nonreproducible results (Hess 1992). Apart from these methods, a few authors (Azzaria 1967; Azzaria and Aftabi 1991; Watling 1981; Peng and Wang 1985) tried to detect different Hg phases in soils and sediments by thermal release techniques. The investigation of different sediments show that Hg release curves do not contradict the results of sequential leaching procedures (Peng and Wang 1985). Azzaria and Aftabi (1991) could distinguish natural Hg-anomalies from polluted soils and sediments by thermal investigations. They concluded that stepwise heating gives more resolution than thermal release curves.

4.2 Idrija

Mining and processing of mercury in Idrija, Slovenia, started in 1490. Nowadays Idrija is one of the biggest Hg-producers of the world. The mined ore, mainly cinnabar, is processed in a roasting plant located near the deposit. Elemental Hg (Hg°) emitted by this roasting plant causes high concentrations of mercury in the atmosphere around Idrija. Recent measurements along the main roads of Idrija showed concentrations up to 100 ng Hg/m^3 air. Even at a distance of 10 km, up to 50 ng Hg/m^3 air were detected (Pirc 1991). Kosta et al. (1974) found amounts up to 570 ng Hg/m^3 air in the city of Idrija and up to 8500 ng Hg/m^3 air near the roasting plant. Previous investigations of Neville (1967) showed values up to 20 000 ng Hg/m^3 air near the deposit. In addition to the geogenic content of Hg in the soils of this area, Hess (1992) supposed an atmospheric influx of Hg into the soils was caused by the emissions of the roasting plant. In order to distinguish geogenic and anthropogenic Hg-loads in these soils, Hg phases were investigated by a temperature-controlled pyrolysis technique (TCP).

4.3 Methods and Materials

The samples used in this work were selected from 147 soil samples of a former investigation made by Hess (1992). His work was focused on the determination of Hg concentrations in the soils of the mining area. The samples were taken in a 127 km^2 sampling grid around Idrija. Additionally, Hg phases were investigated by using sequential leaching procedures.

4.3.1 Temperature-Controlled Pyrolysis

The method is based on the thermal cracking of mercury bonds in solid samples and the release of Hg in its elemental state. Accordingly, different Hg phases are

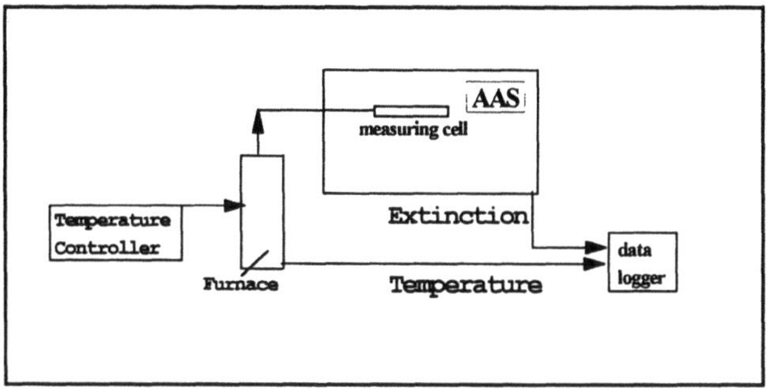

Fig. 4.1. Temperature-Controlled Pyrolysis (TCP)

characterized by different temperature ranges in which Hg is released from the sample. A scheme of the device used is shown in Fig. 4.1.

Up to 200 mg of the dried sample are placed into a quartz vessel. The vessel is inserted into the furnace and the sample is heated continuously in an atmosphere of flowing nitrogen (1 l/min). The temperature of the sample is continuously measured by a thermo-element placed inside the quartz vessel. The Hg released during the heating process passes a measuring-cell and is determined in 1 s-intervals on the 253.7 nm line by cold vapour atomic absorption spectroscopy (CVAAS; *Perkin Elmer AAS 3030*). Extinction and temperature values are continuously recorded by a data-logger. The relationship between temperature and Hg-extinction is expressed by Hg release curves. The temperature ranges in which Hg is released will be named "extinction-ranges" or "peaks".

All samples were measured by a heating rate of 0.5°C/s. Lower rates did not lead to further resolution of additional extinction-ranges, but to a strong widening of the peaks and a rising detection limit. As Hg-free samples show absorption signals due to interferences of non-Hg substances, a background compensator D2 was used to eliminate these interferences.

Natural crystalline cinnabar (Moschellandsberg/Pfalz) mixed with quartz powder was used as a Hg-sulfide (a-HgS) standard. The clay mineral and the iron oxides standard were prepared by exposing the substances for 4 weeks in a sealed container to a Hg° saturated atmosphere. As an example of organic-bound Hg, the Hg-bearing humic acids were extracted directly from the soil samples by leaching the samples 24 h with 1 *M* NaOH at 25°C. The Hg-humic acids (HgHA) were precipitated at pH 2 by adding HCl to the solution. After centrifugation, the humic acids were dried at ambient temperature. The use of synthetic humic acids spiked with Hg-salts caused problems because of the humic substances' property to reduce bivalent Hg to Hg° (Allard and Arsenie 1991). The addition of a solution of $HgCl_2$ to synthetic humic acids always causes a release of Hg° and an additional "Hg°-peak", starting at about 30°C.

Total organic carbon (C_{org}) was determined by photometrically after the combustion of the sample in a high frequency induction furnace (LECO/*CS-225*).

The acidity was determined electrometrically after addition of 125 ml of 0.01 *M* $CaCl_2$ solution to 50 g of the fresh sample.

Total Hg was determined by CVAAS (*Mercury-Analyzer 254 A/Dr. Seitner*) after digesting 5 g of sample material with 28 ml aqua regia for 3 h at 160°C.

4.4 Results and Discussion

The amounts of Hg and C_{org}, as well as the pH values of the samples, are shown in Table 4.1. Mercury concentrations range from 16.8 to 107.6 mg/kg. The content of C_{org} varies between 1.5 and 31.8%, the pH ranges between 5.8 and 7.7.

Table 4.1. Concentration of Hg and C_{org} and pH values of soil samples of the mining area Idrija, Slovenia (data after Hess 1992)

Sample	Hg [mg/kg]	C_{org} [%]	pH
I 42	18.6	1.5	5.8
I 53	31.4	5.7	7.2
I 64	32.0	5.4	7.0
I 73.1	32.0	35.0	6.8
I 73.2	27.0	10.9	7.4
I 74	64.0	2.7	7.4
I 80	17.4	5.2	7.1
I 84.2	107.6	18.8	7.7
I 85	16.8	27.7	6.8
I 113	21.2	4.2	5.8
I 115	23.4	5.1	7.3
I 131	56.2	3.0	6.3
I 143	18.6	15.3	6.5

The Hg-release curves of all samples are shown in Fig. 4.2A. In most samples, all of the Hg is released between 150° and 420°C. All curves show two different extinction-ranges and different ranges of the main release of Hg. The samples I 74 and I 84.2 were chosen as examples for a general description of the two types of Hg-release curves, which were found for all samples. These samples are depicted separately in Fig. 4.2B.

Hg-release from sample I 84.2 starts at about 150°C with a maximum at 250°C for the first peak and at 320°C for the second peak. Most of the Hg is released during the first peak. Mercury-release from I 74 starts at about 165°C with a maximum at 220°C for the first peak and at 330°C for the second peak. In this case, most of the

Hg is released during the second peak. Due to an overlapping of the peaks, the start temperature of the second peak cannot be exactly determined. The presentation of all curves (Fig. 4.2A) shows that the majority of the samples releases most of the Hg during the first peak, in the temperature range between 150° and about 280°C. Only samples I 74 and I 131 show a major release within the second peak, in the range between 280° and about 400°C. In all cases, all Hg was vaporized below 420°C.

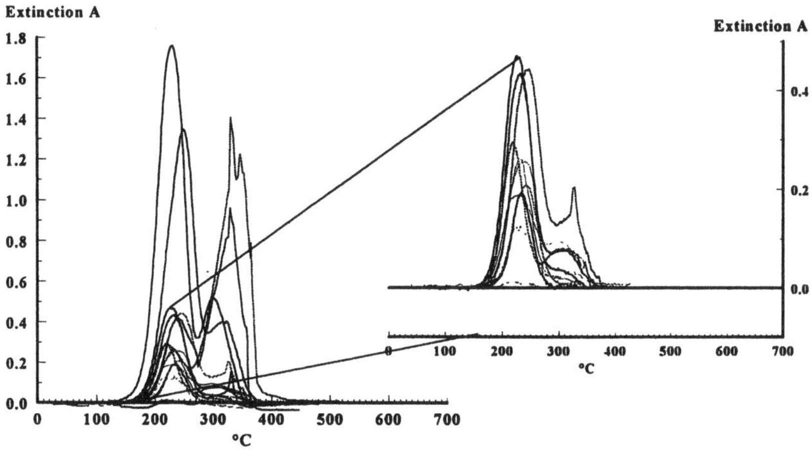

Fig. 4.2A. Hg-release curves of soil samples from the mining area Idrija, Slovenia

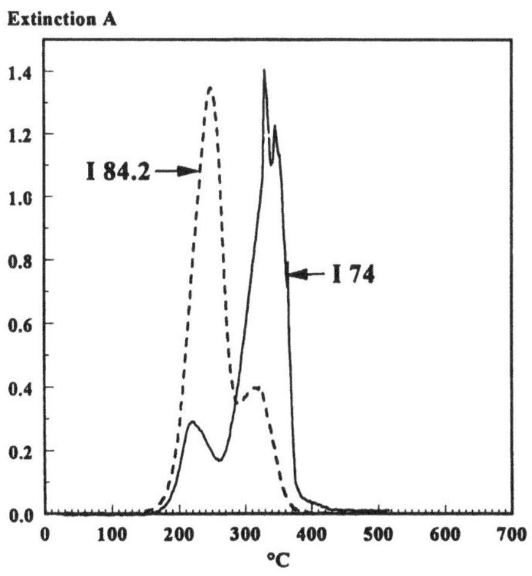

Fig. 4.2B. Hg-release curves of samples I 74 and I 84.2

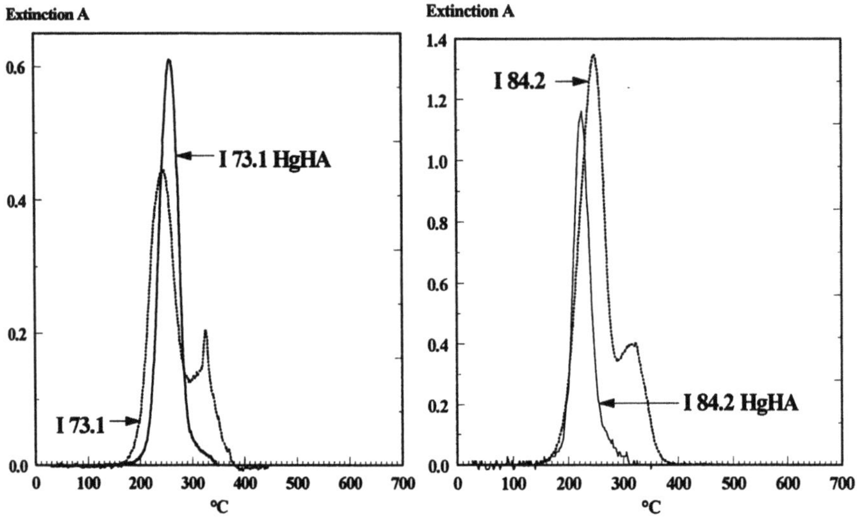

Fig. 4.3. Hg-release curves of samples I 73.1 and I 84.2 compared with their extracted humic acids

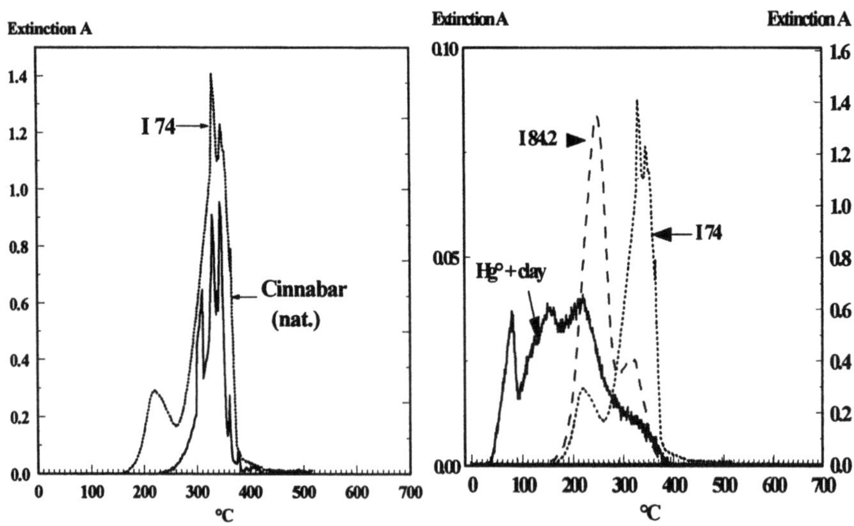

Fig. 4.4. Release curve of the cinnabar standard compared with sample I 74, and release curve of the incubated clay mineral standard compared with samples I 74 and I 84.2.

Mercury phases, determined by TCP, give only information about the relative (thermal) bonding strength of Hg in soils. In order to get more information about how Hg phases are connected with different Hg bonds, the release curves of some selected soil samples were compared with those of some standard samples.

The comparison of the samples I 73.1 and I 84.2 with their extracted humic acids (Fig. 4.3) shows that the Hg-release curves of the humic acids are very similar to the first peak of their original samples. In contrast, the thermal release characteristics of natural crystalline cinnabar correspond to the second peak of sample I 74 (Fig. 4.4). The Hg-release curve of the incubated clay mineral standard (Fig. 4.4) displays a very wide irregular release curve in the temperature range from 30° to about 610°C. The curve rises up to a first maximum at about 110°C and continues with some changes to the maximum extinction at about 240°C.

The curve of the incubated iron oxides (Fig. 4.5) shows two distinguishable extinction ranges. The first range starts at 30°C with a maximum at about 165°C. The transition to the second one is at about 180°C, rising to a maximum at 255°C. The curve of the clay minerals, as well as that of the iron oxides, overlaps the curves of I 74 and I 84.2 in a wide range.

As the starting temperature of the peaks give more resolution for the characterisation of Hg-release curves than the width or the temperature of maximum extinction, the HgHA are in good accordance with the first peak of the soil samples. Nevertheless, in the pH range of the samples (Table 4.1), the sorption of Hg to clay minerals or iron oxides is possible (Anderson 1979). Schuster (1991), however, remarks that organic compounds were more relevant in Hg adsorption at higher Hg concentrations, because the adsorption capacity of organic matter is by far higher than that of mineral compounds.

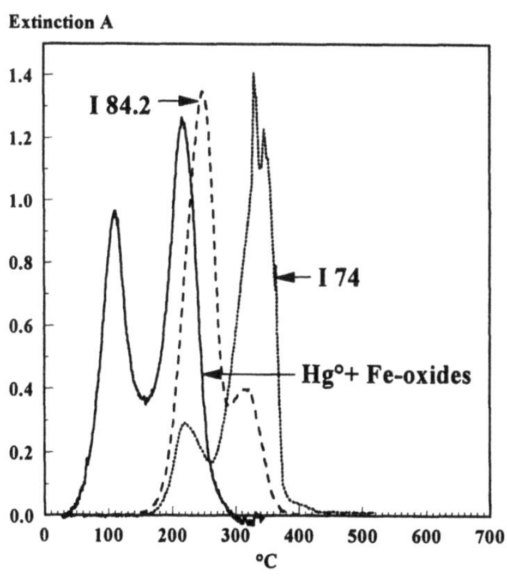

Fig. 4.5. Hg-release curves of incubated Fe-oxides compared with samples I 74 and I 84.2

As cinnabar is the most common Hg species of the Idrija deposit (Mlakar and Drovenik 1971), the occurrence of this mineral, which is quite resistant against weathering, in the soils of the mining area is to be expected. The distribution of cinnabar might be governed by an aeolian transport of dust from the tailings or the use of tailing material for road construction. The association of Hg with humic acids, in contrast, points firstly to the emissions from the roasting plant, and secondly to the deposition by precipitation. Mercury is washed out from the atmosphere and reaches the soil predominantly in a particle-associated form and it is mainly bound to humic substances of the upper soil layer (Lindquist et al.1991; Iverfeldt 1991; Johannsson et al. 1991).

The results of Hess (1992), who applied a modificated sequential leaching procedure after Dües (1987) proved that for all samples more than 90% of the Hg is bound to organic compounds or occurs in the "residual" fraction (Fig. 4.6).

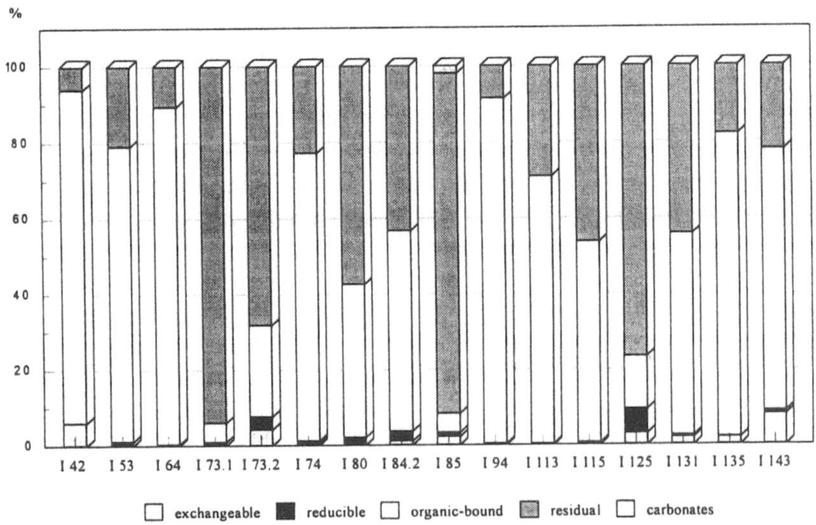

Fig. 4.6. Hg phases determined by a sequential leaching procedure (Hess 1992)

Extractable and reducible forms or Hg associated with carbonates are negligible. Hess (1992) suspected that the "residuals" mainly consist of HgS and Hg°. The existence of free Hg° could not be proven by TCP, however. Our experiments with incubated standard samples revealed that free Hg° in soils is vaporized far below 100°C.

The results of the leaching procedures do not necessarily agree with those of the TCP. The release curve of I 85 (Fig. 4.7) shows that nearly all of the Hg is released during the first peak, while only a small amount of Hg is released during the second peak.

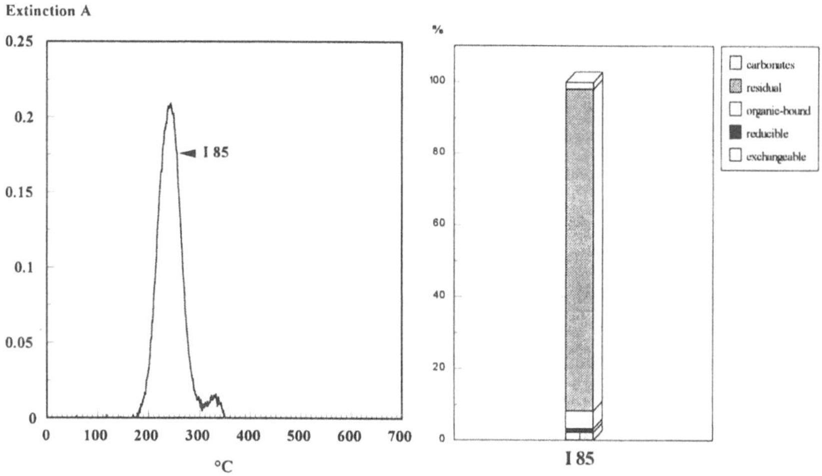

Fig. 4.7. Thermal release curves and results of the extractions by Hess (1992) of I 85

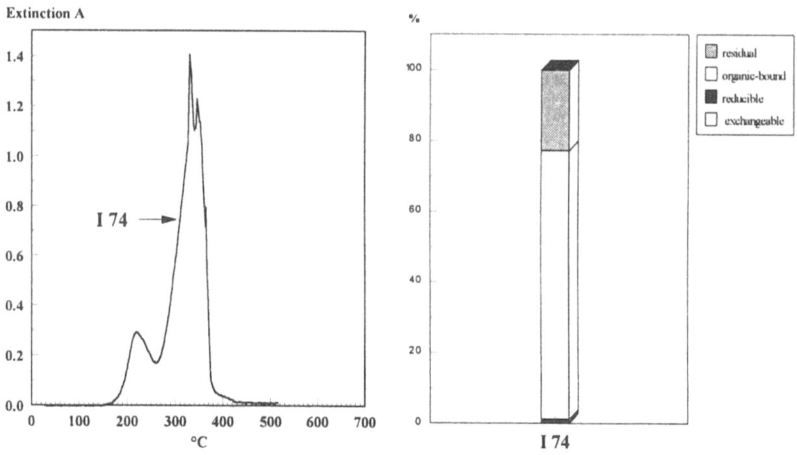

Fig. 4.8. Thermal release curves and results of the extractions by Hess (1992) of I 74

In contrast, the extractions indicate that about 90% of the Hg occurs in the residual fraction and only 5% in the organic-bound fraction.

For the sample I 74 (Fig. 4.8), most of the Hg is released within the second peak, whereas the first peak shows only small amounts. The extractions indicate 23% for the residual and 76% for the organic-bound fraction. The different results obtained by the two methods might be due to an insufficient selectivity of the extractants or insufficient extraction periods.

4.5 Conclusions

The investigation of soil samples in the mining area of Idrija, Slovenia, by temperature-controlled pyrolysis results in Hg-release curves with two different temperature ranges of Hg release. The comparison with standard samples of Hg-incubated clay minerals, iron oxides, extracted humic acids, and of natural cinnabar indicates the occurrence of cinnabar and of Hg bound to humic acids in the soils of this area. Sequential extractions show that most of the Hg occurs in an organic-bound and a "residual" form. The assumption that Hg is preferentially bound to soil organic matter in most of the samples favors an atmospheric influx of Hg, caused by a roasting plant located in the city of Idrija as the main Hg source. In contrast to the immobile cinnabar, Hg bearing humic acids, due to their high mobility in soils, may cause distribution and bioavailability of Hg (Lindquist et al. 1991; Mierle and Ingram 1991).

The TCP is considered to be a suitable method to distinguish between different Hg phases in soils. Compared with sequential leaching methods, TCP shows better resolution of different Hg phases, a shorter duration of the measurements and a good reproducibility of the results.

Acknowledgements. The authors thank Mr. M. Gastner for his technical support in the building and setup of the TCP-device.

References

Allard B, Arsenie I (1991) Abiotic reduction of mercury by humic substances in aquatic system. An important process for the mercury cycle. Water Air Soil Poll. 56:457-464

Anderson A (1979) Mercury in soils. In: Nriagu JW (ed.): The biochemistry of mercury in the environment. Elsevier, Amsterdam, pp 79-112

Azzaria LM, Aftabi A (1991) Stepwise thermal analysis technique for estimating mercury phases in soils and sediments. Water Air Soil Poll 56:203-217

Azzaria LM (1967) A Method of determining traces of mercury in geologic materials. Geol Surv Can, pp 66-54.

Di Giulio RT, Ryan EA (1987) Mercury in soils and clams from a North Carolina peatland. Water Air Soil Poll 33:205-219

Dües G (1987) Untersuchungen zu den Bindungsformen und ökologisch wirksamen Fraktionen ausgewählter toxischer Schwermetalle in ihrer Tiefenverteilung in Hamburger Böden. Hamburger Bodenkundliche Arbeiten 9:265

Glass GE, Sorensen JA, Schmidt KW, Rapp GR, Yap D, Fraser D (1991) Mercury deposition and sources for the upper Great Lakes region. Water Air Soil Poll 56:235-249

Hess A. (1992) Verteilung, Mobilität und Verfügbarkeit von Quecksilber in Böden und Sedimenten am Beispiel zweier hochbelasteter Industriestandorte: Idrija/Slowenien und Frankfurt/M.- Griesheim. Unpubl. PhD. thesis, University of Heidelberg, 171 p

Iverfeldt Å. (1991) Occurrence and turnover of atmospheric mercury over the nordic countries. Water Air Soil Poll 56:251-265

Johannsson K, Aastrup M, Andersson A, Bringmark L, Iverfeldt Å (1991) The coupling of mercury and organic matter in the biogeochemical cycle towards a mechanistic model for the boreal forest zone. Water Air Soil Poll, 56:267-281

Kerndorf H, Schnitzer M (1980) Sorption of metals on humic acid. Geochim Cosmochim Acta, 44:1701-1708

Kosta L, Byrne AR, Zelenko V, Stegnar P, Dermelj M, Ravnik V (1974) Studies on the uptake, distribution and transformations of mercury in living organisms in the Idrija region and comparative areas. Vestnik SDK, 21:49-76

Lindquist O, Johannsson K, Aastrup M, Andersson A, Bringmark L, Hovsenius G, Håkanson L, Iverfeld Å, Meili M, Timm B (1991) Mercury in the Swedish environment - recent research on causes, consequences and corrective methods. Water Air Soil Poll, 55:261

Mierle G, Ingram R (1991) The role of humic substances in the mobilization of mercury from watersheds. Water Air Soil Poll, 56:349-357

Mlakar I, Drovenik M (1971) Structural and genetic particularities of the Idrija mercury ore deposit. Geologija, 14:67-128

Neville GA (1967) Toxicity of mercury vapour. Can Chem Educa, 3, no 1:4-7

Pacyna JM, Münch J (1991) Anthropogenic mercury emission in Europe. Water Air Soil Poll, 56:51-61

Papp C, Filipek LH, Smith, KS (1991) Selectivity and effectiveness of extractants used to release metals associated with organic matter. Appl Geochem, 6:349-353

Peng A, Wang Z (1985) Mercury in river sediments. In: Environ Inorg Chem, VCH Publishers, Inc. Weinheim, 21:393-400

Pirc S (1991) Mercury in the atmosphere of Idrija and surroundings.-(Zivo srebo v ozracju nad Idrijo in okolico). Idrijski razgledi XXXV, 1990/ 1-2, Izdal Mestni muzei Idrija, Idrija, p 1

Revis NW, Osborne TR, Holdsworth G, Hadden C (1989) Distribution of mercury species in soil from a mercury-contaminated site. Water Air Soil Poll, 45:105-113

Schuster E (1991) The behaviour of mercury in the soil with special emphasis on complexation and adsorption processes – a review of the literature. Water Air Soil Poll, 56:667-680

Strohal P, Huljev D (1971) Investigation of mercury-pollutant interaction with humic acids by means of radiotracers. Proc Symp: Nucl Techn Environ Poll, IAEA, Vienna, p 349

Watling RL (1981) The identification and significance of mercury compounds in estuarine sediments. Int Conf Heavy Metals in the Enviroment, CEP Consultants Ltd., Amsterdam, p 591

Part 2

Processes in Soils and Lakes

5 Geochemical Control of Mining Operations

E. Helios Rybicka

5.1 Introduction

Mining activity has been and continues to be a prime contributor to the environmental damages in many countries (e.g. Poland, Canada). Many past mining sites have been heavily polluted by metals, thus, a high quality geochemical database is pertinent to a wide range of environmental studies and should be considered as an essential component of environmental knowledge. The interpretation of the environmental geochemical analysis is very complicated by the variety of sources, the kind of pollutions and the geochemical processes (e.g. early diagenesis processes) as well.

Of the many mineral commodities exploited for thousands of years as well as recently, the most important are: hard and brown coals, iron, chromium, copper, zinc, lead and nickel ores, sulphur and rock-salt. Very often the metallurgical works are situated near the metal mines, and thermal power and heating plants are found in the vicinity of coal plants, thus, geochemical factors prevailing in mining areas seem to be complex. The major environmental impacts of mining and metal beneficiation industries are the following: (1) changes in hydrogeological systems, (2) hydrological transformation of soils and surficial flows, (3) contamination of soils and surficial water reservoirs and (4) chemical pollution of the atmosphere. In order to show the magnitude of this last problem, in 1991 the total amount of dust particulate emission in Poland was 1950 tons, from which 1177.8 tons was caused by power plants and 151 tons by metallurgical processing.

The hydrological system of mining areas is influenced mainly by mineralised and polluted waters originating from the dewatering of mines and discharging to the surficial flows. These waters are rich in Cl^- and SO_4^{2-} ions and contain a high amount of heavy metals and natural radioactive isotopes (uranium, radium) as well.

An important problem is the contamination of both the bottom and flood-plain sediments by heavy metals derived from mine waters as well as released by ore processing and smelting plants. In Poland the highest concentration of Zn (11 580 ppm), Pb (1745 ppm) and Cd (200 ppm) were reported from the Przemsza River sediments.

Mines produce tremendous quantities of waste rock and tailings which have to be disposed of on land or into aquatic ecosystems. The increase of waste material is depending on the type of exploitation, the quality of raw material output as well as on the efficiency of beneficiation. It is estimated that in Poland during the years 1984–2000 about 900 million cubic metres of spoils will be dumped only in the area of the Upper Silesian Coal Basin. The annual load of chlorides in these spoils reaches 40 000 tons.

Soils in the neighbourhood of metal mines and smelters show metal concentrations exceeding up to 100 times the background. In Pb–Zn mining regions of Poland, the highest concentrations of some metals are found, e.g. Pb 8200 ppm, Cd 140 ppm, Zn 13 250 ppm. Also very high amounts of Cu up to 7400 ppm and Pb 2000 ppm were found in soils near copper mines and smelters.

5.2 General Processes Providing Metallic Pollution

The following kinds of ore mining and processing are mainly responsible for metallic contamination in the environment: (1) iron ores, (2) non-iron ores e.g. zinc, lead, copper, and (3) hard coal. The accumulation of metals in different environmental compartments i.e. ground waters, surficial flows and soils, can be the result of sulphide oxidation taking place in both mineral deposits and deriving from mining and processing residues. The reaction processes can be described in general as follows:

(1) $PbS + 2O_2 \rightarrow Pb\,SO_4$
galena

(2) $CuFeS_2 + 4O_2 \rightarrow Cu\,SO_4 + Fe\,SO_4$
chalcopyrite

(3) $2FeS_2 + 7O_2 + 2H_2O \rightarrow 2FeSO_4 + 2H_2SO_4$
pyrite
$4FeSO_4 + 2H_2SO_4 + O_2 \rightarrow 2Fe_2(SO_4)_3 + 2H_2O$

These processes may cause a pH decrease to about 3 in the infiltrating waters. These types of processes take place usually in weathering zones of natural and anthropogenic sulphide deposits leading to the sulphate enrichment of ground waters. The same situation (a pH decrease and sulphate enrichment of aquatic systems) appears in case of the weathering of native sulphur deposits.

Within the oxidation zones of copper-pyrite deposits very complicated processes take place due to the activity of micro-organisms. The processes can be described by the following reactions:

$$CuFeS_2 + 4O_2 \rightarrow CuSO_4 + FeSO_4$$

$$4CuFeS_2 + 2Fe_2(SO_4)_3 + 2H_2O + 3O_2 \rightarrow CuSO_4 + 5FeSO_4 + 2H_2SO_4$$

The produced sulphuric acid facilitates the solution of Cu in waters and its migration in the form of sulphates or other compounds like organic and inorganic complexes, especially chelates. The migration of Cu in aquatic systems is hindered by the presence of $CaCO_3$, by adsorption on hydrous Fe-Mn-oxides, clay minerals and organic substances. In addition to this, the presence of H_2S in extreme situations may reduce completely the aureole of waters enriched in Cu around the deposit.

Similar reactions can take place by the weathering of Pb sulphide deposits, although the precipitated anglesite ($PbSO_4$) acts as a protection against the further penetration of waters rich in oxygen, a process which slows down the reaction.

The sequential character of weathering of zinc sulphide deposits can be presented as follows:

Step I: oxidation processes within the deposit:

$$3ZnS + 4Fe_2(SO_4)_3 + 2H_2O + 4O_2 \rightarrow 3ZnSO_4 + 8FeSO_4 + 4H_2SO_4$$

In the presence of pyrite the process becomes more intensive.

Step II: processes in the vicinity of deposits within calcareous rocks:

$$CaCO_3 + ZnSO_4 \rightarrow ZnCO_3 + Ca\,SO_4$$

Step III: processes in the vicinity of the deposit as a result of acidic effluent water formed during step I:

$$ZnCO_3 + H_2SO_4 \rightarrow ZnSO_4 + H_2O + CO_2$$

The processing of metal ores introduces a variety of chemicals to the mine waste tides including metals also from non-ore fractions, nitrogen compounds from blasting operations, colloidal suspended solids, and thiosalts ($S_2O_3^{2-}$; $S_3O_6^{2-}$; $S_4O_6^{2-}$), which on further oxidation cause acidification of water systems.

5.3 Contaminated Rivers

The water and sediment of many rivers and estuaries which drain mining and metal processing areas e.g. in south-west England and southern Poland, have been found to have high heavy metal contents (Aston and Thornton 1973; Johnson and Thornton 1987; Macklin and Klimek 1992; Helios Rybicka 1992). The rivers drain a mineralised area consisting of rocks of different ages as well as petrographical types, which have been intruded by mineral lodes containing main sulphide ores, like FeS_2, $CuFeS_2$, ZnS, PbS, and traces of other sulphised ores, such as CdS, HgS, TlS.

In many cases, the considered areas have a long mining history. They are very degraded, with a sparse vegetation cover. The oxidation of sulphide minerals has produced highly acidifying mine waters with a high trace metal content. The additional load of polluted waters which originates from the dewatering of mines is supplied to the rivers as well. These waters usually contain high concentrations of chlorides and sulphates, thus, some metals may be transported not only as simple aquo-ions but also as soluble labile complexes of Cl^- and SO_4^{2-}. The forms in which the trace metals are transported depend on their chemical properties and their interaction with particulate material.

In order to illustrate the major geochemical factors controlling the impact of mining activities on the environment, it is necessary to investigate the chemical form of the contaminant, e.g. their chemical speciation in both river water and sediment. The speciation can help to assess the bioavailability of metals to organisms and the hazard they may present to the environment.

The results of a detailed investigation of the Carnon River system (south-west England) have shown high concentrations of Fe, Cu, Zn, As and Cl^- in the river water (Johnson and Thornton 1987), with the Cl^- subjected to strong seasonal variations. Over 80% of Fe, Zn and As, and about 40% of Cu is derived from acid mine drainage. The binding of Cu and Zn depends on the pH and Fe concentration, whereas, only little pH dependence is observed for As. About 80% of arsenic was found in the particulate phase.

River water from the mining areas contain a high amount of iron which is transported largely in a particulate form. This material is providing an active surface for adsorption or/and coprecipitation processes, which are responsible for the final distribution of heavy metals between the dissolved and particulate form.

In regions with a high mining activity, the river catchments with a long history of metal extraction and manufacture, are the areas of severe heavy metal pollution, and of many other environmental problems. During this long-term contamination, pollutants – particularly heavy metals – are accumulated by the solid phase and concentrate in the overbank alluvium within the river basin, as well as in bottom sediments. Therefore, aquatic sediments may better characterise the degree of the environmental impact in a certain mining areas. To stress the importance of this problem, one has to realise that overbank alluvial sediments once polluted by past and present mining and other industrial activities, constitute the most important secondary source of metal contaminants.

Recent investigations (Macklin and Klimek 1992) show very high concentration of Zn, Pb and Cd in the overbank alluvial sediments of the Upper Vistula (Metkow) and Przemsza (Gorzow) Rivers. Cadmium concentrations are particularly elevated and approach some of the highest levels (up to 150 ppm) recorded in river sediments in Europe.

Also, the content of heavy metals in bottom sediments of these two rivers is very high. In the Przemsza River it can reach : 11 664 ppm Zn, 868 ppm Pb and 495 ppm Cd, and in Vistula River: 3000 ppm Zn, 450 ppm Pb and 100 ppm Cd.

The metal concentrations in the suspended matter are even higher, e.g. the Vistula River can obtain about 6000 ppm Zn, 1200 ppm Pb and 134 ppm Cd (Helios Rybicka 1992).

5.4 Contaminated Ground Waters

The hydrological system in mining areas is influenced by mineralised and polluted waters, which originate from mine drainage. High concentrations of sulphates, chlorides and trace metals in ground water results from the infiltration of industrially contaminated meteoric waters, from leaching of numerous spoil dumps and from the long-lasting residence of ground water in abandoned workings.

To obtain reliable information about the degree of contamination of migrated water, it is necessary to study the pore-water solution and leachate composition within the surface layer of mining wastes. For that purpose the concentrations of some trace metals, SO_4^{2-} and pH were measured during an 8-year period (1985–1992) in the leachate from the surface layer of fresh wrought and from 10-year-old coal wastes taken from Smolnica tip (Patrzalek et al. 1994). The results show a relatively high sulphide reactivity and a low buffering capacity, which means an insufficient content of Ca-Mg carbonates in the solid matrix and a rather high susceptibility to acidification. The quality of waters flowing to the tip and adjacent areas corresponds to standards for potable waters, whereas, concentrations in the outflows exceed the standards even for the lowest, third-quality class for inland waters.

It must be emphasised that percolating water from tailing ponds are highly contaminated and depending on contaminants, may affect the ground water. For example in the Lubin Copper District in Poland, the water from tailing ponds contains high amounts of Cu, Zn and Pb sulphates and chlorides as well as organic compounds. If these soluble metal compounds contaminate the receiving ground water, its use as potable water is restricted and may also be harmful for vegetation (irrigation).

5.5 Contaminated Soils

The major factors affecting the quality of soils close to mining industries are: (i) change of soil pH, i.e. acidification or alkalisation, as well as (ii) accumulation of trace elements, particularly of heavy metals. Because those soils are contaminated mainly by atmospheric depositions, it is difficult to specify the sources in highly industrialized areas. As an example, most of Poland's industry is concentrated in the Katowice district, including 55 coal mines, 23 power plants, 14 steelworks and smelters, 2 non-ferrous metal works, 4 non-ferrous metal ore mines. About 50% of the country's steel production, 98% of coal and 100% of Zn-Pb ore output is localised within this heavy-industry area. About 4000 industrial plants operating in this district have been classified as hazardous to the environment. Nevertheless, agricultural land makes up to 47.1% of the Katowice district area covering about 313 000 hectares.

Table 5.1. Soil contamination in the Upper Silesia Region, in mg/kg

Site	Pb	Cd	Zn
Bytom	129–2290	2– 85	193–12592
Bukowno	46–1520	1– 42	90– 9200
Tarnowskie Góry	26–8200	1–143	103–13250
Katowice	20–1050	n.d.– 20	61– 2110
Chrzanów	24–1100	1– 35	62– 5660

Table 5.1 shows the range of zinc, lead and cadmium concentrations in soils of various parts of the Katowice District. In extreme cases, the Pb concentration within the arable layer of soil amounted to 8000 mg/kg, Cd to about 150 mg/kg and Zn up to 113 000 mg/kg (Michna and Zurek 1991 from Kucharski et al. 1992).

The question is, whether such a region where mining, industrial and agricultural activities are overlapping, can be a good test site to specify the geochemical factors controlling mining operations.

5.6 Final Remarks

Most heavy metals as well acids in mining areas originate from the weathering of sulphides (e.g. FeS_2, PbS, ZnS) contained in spoils (e.g. in the Upper Silesian Coal Basin under full oxygenation, the decomposition of sulphides may yield per year 450 000 tons of sulphates). Many dumps may produce acid mine water with a pH below 4.

A combined hydrological, chemical and/or geochemical approach can provide much valuable information about critical concentrations of trace metals, when estimating the environmental impact of mining and processing activities on river, ground water and soil systems. This information makes it possible to define the chemical factors which determine the dissolved and particulate portions of metals. Besides organic matter, both clay minerals in the colloidal form as well as amorphous iron and manganese oxides/hydroxides, which together can make up to more than 12 weight-% of aquatic sediments, are important scavengers of heavy metals due to their (1) large specific surface area, (2) high negative surface charge and (3) high cation exchange capacity. These solids may act as metal "donors" at contaminated sites, and under specific ambient conditions (pH, Eh, ionic strength) provide the formation of new metal phases. Under anoxic sediment conditions, most heavy metals should precipitate as rather insoluble sulphides, e.g. ZnS, HgS, CuS, PbS and FeS_2 (Helios Rybicka 1992).

These considerations of the general behaviour of metals in aquatic and soil systems should be used to assess and control accumulation and mobilisation processes of trace metal ions. To evaluate the environmental impact of mining activities with regard to chemical contaminants, the following main steps of geochemical activities are recommended: (1) estimate the extent of the local and regional contamination, (2) quantify the sources of heavy metals, chlorides, sulphates and thus the total acidity and salinity, as well as other contaminants and (3) define and control major sources, i.e. waste rock heaps, tailings, waste dumps and air pollution, etc.

References

Aston SR, Thornton I (1973) The application of regional geochemical reconnaissance surveys in the assessment of water quality and estuarine pollution. Wat. Res. 9.

Helios Rybicka E (1992) Phase-specific bonding of heavy metals in sediments of the Vistula River, Poland. Appl. Geoch. 7.

Johnson CA, Thornton I (1987) Hydrological and chemical factors controlling the concentrations of Fe, Cu, Zn and As in a river system contaminated by acid mine drainage. Wat. Res. 21: 3.

Kucharski R, Marchwinska E, Gzyl J (1992) Agricultural policy in polluted areas. Sci. Tot. Env.

Macklin MG, Klimek K (1992) Dispersal, storage and transformation of metal-contaminated alluvium in the upper Vistula basin, Southwest Poland. Appl. Geoch. 12.

Patrzalek A, Twardowska I, Szczepańska J (1994) Biological reclamation of coal mining waste tip as an essential factor in its resultant environmental impact. (In press).

6 Chemical Speciation and Fate of Selected Heavy Metals in Soils Strongly Polluted by Copper Smelters

A. Karczewska

6.1 Introduction

In Poland, mining and smelting of non-ferrous metals are two important sources of heavy metals in the environment. Being one of the most important European centres for copper metallurgy, LGOM (the copper industry region of Legnica and Głogów, situated in the south-western part of Poland) contributes to environmental contamination by releasing phytotoxic gases and metallurgical dusts into the air, and to the surrounding soils and vegetation. Over the last forty years, three copper smelters have discharged large amounts of gaseous pollutants and heavy metal containing dust (especially copper and lead, but also zinc, cadmium, arsenic and manganese). Dust depositions and the direct influence of phytotoxic gases (in particular of sulphur dioxide) on plants caused degradation of agricultural systems, including soils, not only close to the smelters, but also over a distance of more than 10 km from the emission source, where adverse effects on the crop yield and its quality were observed. The degree of soil degradation depended on the distance from the smelter and also on other factors, such as topography, climatic conditions and, of course, on particular soil properties.

Since the early 1970s, the Institute of Soil Science and Agricultural Environment Protection of the Agricultural University of Wrocław has carried out monitoring of the heavy metal content in soils, in order to estimate environmental risks associated with soil contamination (Kowaliński et al. 1972, 1979; Roszyk and Szerszeń 1988ab; Szerszeń et al. 1993; Borkowski et al. 1991). Some results and conclusions of this work have been briefly summarised and reported in this chapter. However, the main theses of Chapter 6 refers to the mobility and bioavailability of metals in contaminated soils, as the main aspects of the observed soil pollution. Many papers have already been published on this subject (Adriano 1986; Alloway 1986; Gorlach and Gambuś 1991ab; Hornburg and Bruemmer 1993; Kabata-Pendias and Pendias 1993; Mc Bride 1989), some of these with particular reference to the situation found around the copper industry (Komisarek 1990; Rachwal 1990; Kabata-Pendias 1979; Karczewska 1995), but without giving any final coherent idea on the fate of heavy metals in polluted soils. Depending on the research method, some authors set forth arguments bearing out a definite metal immobilisation (Szerszeń et al. 1991; Shamsham 1992, 1993), whereas others emphasized the metal mobilisation potential, even at relatively "safe" soil conditions, i.e. at neutral reaction and high CEC values (Kociałkowski et al. 1991; Greinert 1993; Karczewska 1995).

6.2 Objective of the Research Work

Estimating the potential risk of heavy metal mobilisation from the strongly polluted soils surrounding copper smelters was the main objective of this work. Various approaches have been made to describe the fate of heavy metals from metallurgical sources during their long-term occurrence in the soil compartment. Although metals are mainly accumulated in the surface layer, without any clear evidence of increasing Cu or Pb concentrations in the subsoil, the risk of metal mobilisation and downward transport through the soil profile and into the groundwater is of prime environmental concern and should be thoroughly examined.

As it is the chemical form rather than the total concentration of a metal pollutant in the soil, which determines the environmental availability, research was carried out to describe the main association forms of metals in strongly polluted soils, and to examine their transformation under different soil conditions.

6.3 Smelters and Soils: General Characteristics

Since 1954, when the first copper smelter in LGOM, "Legnica", was opened, the copper industry in this region has been successively developed. Two other plants "Głogów I" and "Głogów II" were opened in 1971 and 1977 very close to each other and in a distance of about 40 km from the first (Fig. 6.1). Now, these three smelters produce about 400 000 tons of copper yearly and at the same time release more than $2 \cdot 10^5$ tons of gaseous pollutants (SO_2, CO, NO_x, CS_2 and others), and more than 2600 tons of metallurgical dust (containing about 200 tons of copper, 150 tons of lead and other heavy metals) into the atmosphere. In the late 1970s and early 1980s, the annual emissions of the smelters used to be more than 10 times higher than it is now (see Table 6.1), in spite of the much lower copper production in the past. A significant decrease of emissions from the smelter operations in the late 1980s, was the result of modernisation measures, focusing on the maximum possible reduction of their environmental impact. Nevertheless, the surrounding areas still remain strongly contaminated with heavy metals, which had accumulated in the soils during many years. Current depositions, although relatively low, contribute to the already existing pollution.

Table 6.1. Emissions of the most hazardous gases and dusts from copper smelters Legnica and Głogów, tons per year (1971–94)

Pollutant	Emitter	Yearly emmission			
		1971–83 [a]	1984	1989	1994
Dust (total)	Legnica	15200	3800	1250	1050
	Głogów I,II	18750	3620	1860	1630
Copper	Legnica	960	205	70	70
	Głogów I,II	1640	215	130	115
Lead	Legnica	1720	320	61	70
	Głogów I,II	1330	430	114	70
Sulphur dioxide	Legnica	27000	21600	12600	4350
	Głogów I,II	125700	94200	37600	27500
Sulphuric acid	Legnica	360	270	120	120
(mist)	Głogów I,II	2750	2150	970	350

[a]maximum

Soils in the area affected by smelting activities differ in their genesis, texture, chemical, physical and physico-chemical properties, and belong to several taxonomic units. In the vicinity of the smelter "Legnica" there are brown and lessives soils (Eutrochrepts, Dystrochrepts and Hapludalfs according to the US Soil Taxonomy 1975) developed from alluvial loams and silty loams on underlying sand and gravels. Owing to good physical and physico-chemical soil properties, most of the area near the smelter "Legnica" was used as arable lands before setting up the smelter. In the neighbourhood of smelters "Głogów", soils are much more differentiated. In the southern part of this area, which is slightly hilled (Dalkowskie Hills), soils are similar to those near "Legnica". North of the smelter, on both sides of the Odra river, there are predominantly river alluvial soils (Udifluvents, Fluvaquents and Humaquepts) developed from parent materials representing a wide spectrum of various textures, both extremely light (sands con-

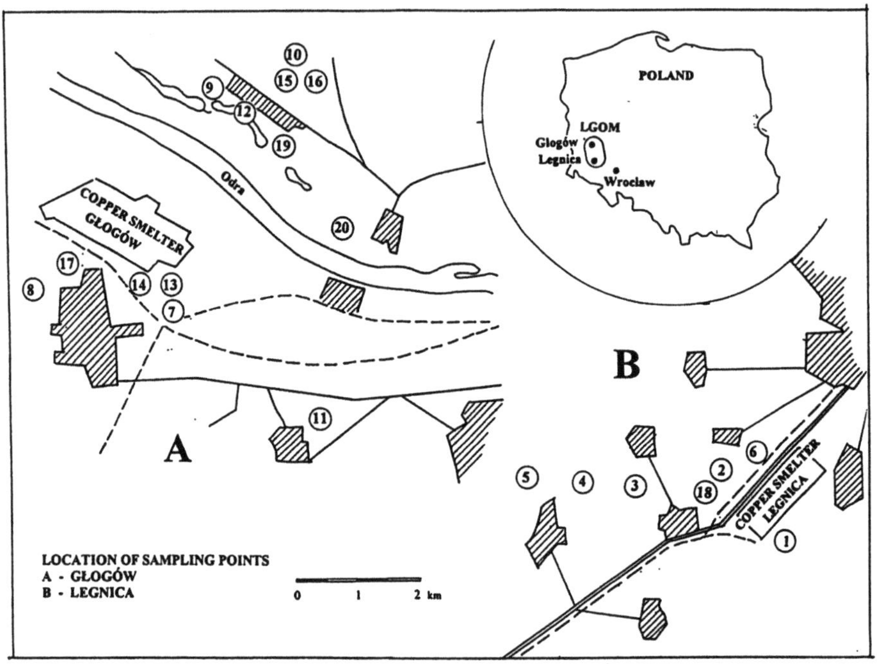

Fig. 6.1. Locality of smelters and soil sampling points

taining less than 3% of clay fraction) and heavy (silty clays, loams and clay loams of the clay contents as high as 40–50%; see Table 6.2). Most of the soils surrounding of smelters "Głogów" were previously used as arable lands and only within the reach of river floods as grasslands. Because of the high level of air and soil pollution, two large "sanitary protection zones" have been created around the smelters, each of about 4000 ha, and former fields are no longer agriculturally used. The area has been bought up by the copper company, which also took over the responsibility for soil reclamation and underground water protection. At first, soils within zones were limed to cause heavy metal immobilisation. The area is now being partly forested, predominantly with poplars which turned out to be most resistant to air and soil pollution, and to give the highest possible yields of wood to be utilised in the smelting process. There are some experimental fields left within the sanitary protection zones, where experiments are being carried out in order to find and optimise options of land reclamation. Some of them are focused on growing non-consumable plants, such as willow, flax or hemp. Special attention has also been put on the monitoring of heavy metal concentrations in soils, the mobility and bioavailability of metals, and possible phytotoxic effects on cultivated plants.

Table 6.2. Total content of selected metals in the surface samples (0–20 cm) of arable or post -arable soils in years 1988–93

Smelter	Distance, km Direction	Location No.	Clay %	Textural group	pH 1*M* KCl	CEC (mean) mval/100g	C_{org} (mean) %	Total content of metal, mg/kg Cu	Pb	Zn	Cd
Legnica	0.5 SE	1 [a]	2–4	sandy loam	6.0–7.0	7.5	1.2	191–245	90–108	55–96	0.32–0.54
	1.0 W	2	3–5	silty loam	6.3–7.2	11.3	1.5	750–986	297–360	121–165	1.32–1.63
	2.0 W	3	3–5	silty loam	6.2–6.6	8.9	1.4	250–280	114–170	68–90	0.42–0.58
	3.0 W	4	4–8	loam	6.7–7.2	10.1	1.2	100–248	67–130	59–80	0.38–0.49
	4.0 W	5	4–7	silty loam	5.8–6.6	9.4	1.5	75–101	45–130	44–80	0.20–0.36
	0.5 NW	6 [a]	8–12	silty loam	6.6–7.1	9.5	1.1	480– 519	220–296	150–186	1.53–1.70
		according to Roszyk and Szerszen (1988 a,b)					range:	25–9800	30–2200	92–2370	0.05–2.44
		within sanitary protection zone (278 samples) [b]					mean:	215	130	183	0.42
Głogów	1.0 SE	7 [a]	4–8	silty loam	6.8–7.3	8.4	1.1	250–317	88–130	42–55	0.39–0.51
	1.0 SW	8 [a]	2–5	sand	7.1–7.4	7.0	0.9	165–325	50–130	42–65	0.28–0.35
	3.0 NE	9	1–6	sand	7.2–7.5	3.6	0.5	398–435	122–170	36–100	0.55–0.67
	4.0 NE	10	25–36	(clay) loam	6.3–6.9	43.5	2.6	255–306	111–265	92–160	0.60–0.68
	4.0 SE	11	3–6	silty loam	5.9–6.3	6.9	1.3	78–100	31–90	36–50	0.12–0.19
	0.5 SE	13	2–6	sandy loam	5.2–6.5	4.9	1.3	880–1840	240–480	68–98	0.35–0.78
	0.1–5.0	according to Roszyk and Szerszen (1988 a,b)					range:	30–3280	30–18400	25–4000	0.02–10.90
		within sanitary protection zone (529 samples) [b]					mean:	460	219	124	1.06

[a] after a deep plough and liming carried out in 1980s
[b] including grasslands

6.4 Material and Methods

Several approaches were made to examine and describe the fate of heavy metals in polluted soils. Some parts of this study, focused on metal concentrations in soils, were carried out in the field and other parts – referring to metal forms and their transformations – in pot and laboratory experiments.

The field approach involved a regular monitoring of heavy metal content in soils at 12 fixed sampling points situated at different distances of 0.5 to 4.0 km from the sources (Fig. 6.1, points 1–11 and 13). The sampling points were chosen thoroughly so that they would cover the wide spectrum of soil properties and of total metal concentrations. Soil samples were taken from two depths at 0–20 cm and 30–40 cm. At some selected sampling points (described in following), the whole soil profile down to 150 cm, was also examined. These profiles were chosen so as to be representative for a given part of the area.

Soil and plant samples were collected from the field in three replicates which were either analysed separately or mixed to obtain the average samples. All laboratory analyses were carried out in three replicates and the results given in this chapter represent mean values.

Basic analytical procedures followed the Polish standard methods of soil and plant analyses. Soil samples were dried at room temperature, crushed in a porcelain mortar and sieved through a 1-mm screen. Soil texture (particle size distribution) was determined using the modified sedimentation method of Cassagrande and soil textural groups were defined according to the ISSS classification. Basic chemical and physico-chemical soil analysis involved: pH (determined in 1*M* KCl), organic matter content (oxidometric analysis according to Tiurin), cation exchange capacity (according to Kappen and Pallman) and analysis of selected heavy metals, i.e. Cu, Pb, Zn, Cd and also Fe and Mn. Total content of metals in soil samples was determined after digestion with concentrated HNO_3 + $HClO_4$, metal content in plant material after the dry thermal mineralisation followed by dissolving the ash in HNO_3 + HCl. Heavy metal concentrations in solutions were measured by flame atomic absorption spectrophotometry. More specific methods used in this research will be described later on.

6.5 Total Content and Profile Distribution of Heavy Metals in the Affected Soils

Research work on the heavy metal load of soils from the area affected by copper smelters has been carried out since the early 1970s, i.e. since the smelter "Głogów I" began its production. A great deal of research has been done on this subject over the past several years, generally confirming the high level of soil contamination found for copper and lead. Also elevated concentrations of some other metals, such as zinc, cadmium and arsenic, have been reported. Metal concentrations in soils depended strongly on the distance from the emission source and on wind direction, with the most frequent wind directions between south-west and north-east (Kowaliński et al. 1979; Kabata-Pendias et al. 1981ab; Roszyk and Szerszeń 1988ab; Szerszeń et al. 1993). Since heavy metals remain rather immobile in soils, their highest concentration in soils affected by dust depositions occur close to the soil surface. The maximum copper and lead concentrations reported in the surface soil horizons, were found in a distance less than 1 km from the smelters and amounted at 9800 mg/kg Cu and 4580 mg/kg Pb near "Legnica", and 5000 mg/kg Cu and 18 400 mg/kg Pb near "Głogów I, II" (Roszyk and Szerszeń 1988ab; Weber 1987).

Table 6.2 shows the total metal content in the surface soil layer, in relation to the distance from the smelter, and based on our own work and on other data. Relatively low Cu and Pb concentrations measured close to emission sources probably resulted from the deep ploughing undertaken as a first step of soil reclamation. In this way, the heavy metals previously accumulated in the surface layer were "diluted" in the top-soil. It is not definitely known whether such treatment would, or would not, cause any changes in metal mobility.

From the early 1970s until mid-1980s, when the emissions were radically cut down, the continuous increase on Cu and Pb concentrations in the surface layer of arable soils was observed at all sampling points examined. This increase was statistically confirmed by linear regression. The most intensive Cu and Pb accumulation in soils took place at points which were situated nearest to the smelters (points 2 and 13). A strong accumulation occurred first of all in the surface soil layer and there was no proof that the metals migrated to deeper soil strata. Their concentrations in the subsoil layer (30–40 cm) did not show any significant changes over more than 20 years, though concentrations were higher than the geochemical background in similar parent rocks (see following page Table 6.3; Czarnowska and Gworek 1991; Dudka 1992; Kabata-Pendias and Pendias 1993). Concentrations of Zn, which was the third metal under investigation, did not increase significantly in either of the two soil layers, but rather decreased in most cases. This long-term observation has continued for recent years, although deep ploughing, carried out in some parts of sanitary protection zones disturbed the natural arrangement of soil horizons and caused metal "dilution", as described above. The latest results obtained from undisturbed soils showed, that total Cu and Pb contents in the surface soil layer remained at a stable level or slightly decreased. Figure 6.2 shows the changes of Cu, Pb and Zn concentrations in the surface and subsurface soil layers at two selected sampling points, i.e. 7 and 10.

Table 6.3. Background concentrations of selected heavy metals in Polish non-polluted soils

Soil type and texture	Element	Soil horizon	Metal concentrations, mg/kg Range	Mean
Alluvial	Cu	A	16–28	22
various textures		B	14–25	20
	Pb	A	13–49	39
		B	(2–20)	
	Zn	A	55–125	85
	Cd	A	0.24–0.36	0.3
	Mn	A	150–1965	1085
Lessives	Cu		8–54	19
silty soils	Pb		14–32	26
	Zn	A	20–130	50
	Cd			0.38
	Mn		110–1060	470
Brown	Cu		4–36	16
loams and	Pb			22
silty soils	Zn	A	30–360	65
	Cd			0.45
	Mn		50–1400	500
Podzols	Cu		1–26	8
and other	Pb		8.5–23.5	16
sandy soils	Zn	A	7–150	30
	Cd		0.01–0.24	0.07
	Mn		15–1500	240

The analysis of results obtained from this research did not give an answer whether or not the heavy metals from metallurgical sources migrate downwards in the soil profile. In order to examine all possible evidence for or against such an elemental migration, selected undisturbed soil profiles, down to a depth of 150 cm, were analysed for heavy metals in relation to other soil properties. Some results, representative for all profiles examined, are illustrated in Fig. 6.3. It can be easily seen that a strong accumulation of metals occurred in upper horizons, close to the

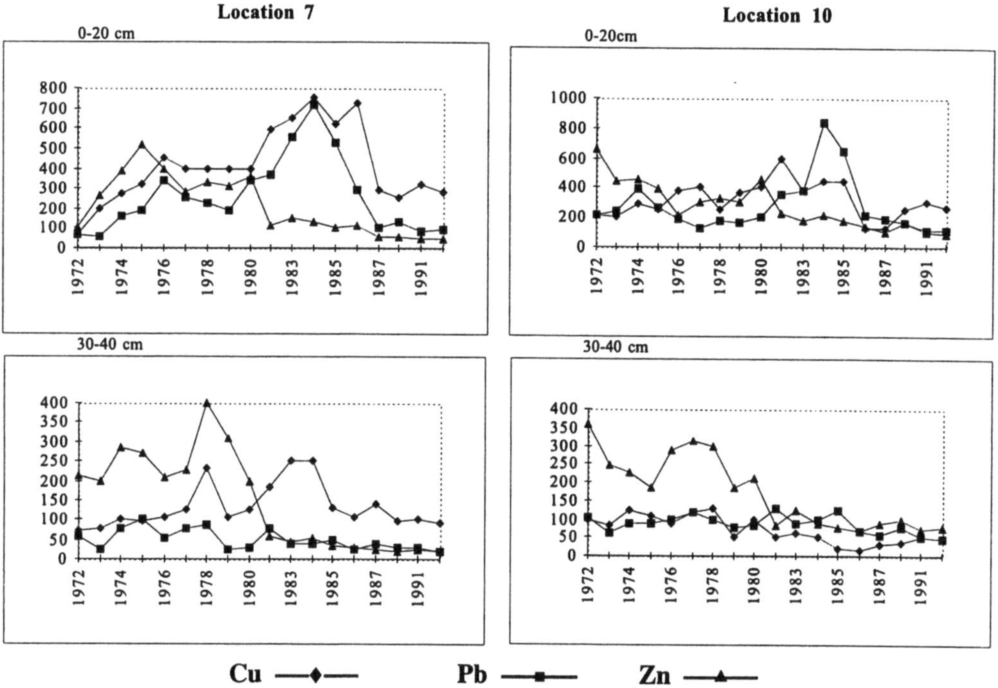

Fig. 6.2. Changes in total metal concentration, mg/kg, in surface (0–20cm) and subsurface (30–40 cm) soil layers in selected soils during 1972–93

soil surface, depending on the way of soil application within a layer of about 0–20 cm in arable lands, and within 0–5 (0–10) cm in grasslands.

Metal concentrations in deeper soil horizons were relatively low and no evidence was found for metal leaching from the upper soil horizon into deeper layers. However, there are some indications that some mobilisation might have happened. Although no significant enrichment of heavy metals occurred within soil profiles, the metal concentrations measured in the horizon below 50 cm were slightly (but significantly at the $p=0.05$ level) higher than the background for the similar rocks. Thus, metals could have been leached. These may remain mobile, however, because of the low sorptive capacity and corresponding properties of deeper soil layers.

The "field approach" based on metal distribution within soil profiles did not give a clear answer whether or not heavy metals were leached downwards from the surface layers. The negative result of long-term field monitoring prompted a new approach to the problem, i.e. the determination of the chemical forms of heavy metals in soils and their changes under controlled conditions.

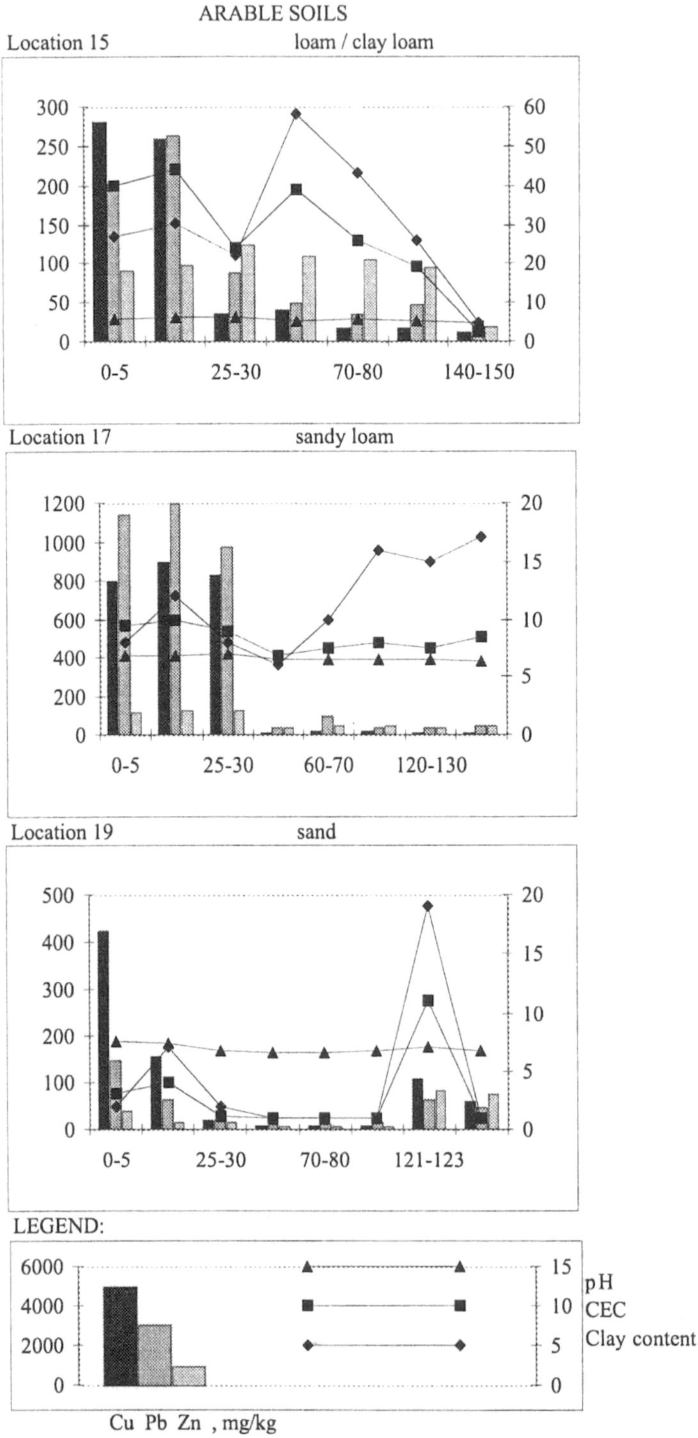

Fig. 6.3. Profile distribution of Cu, Pb and Zn in soils near copper smelters

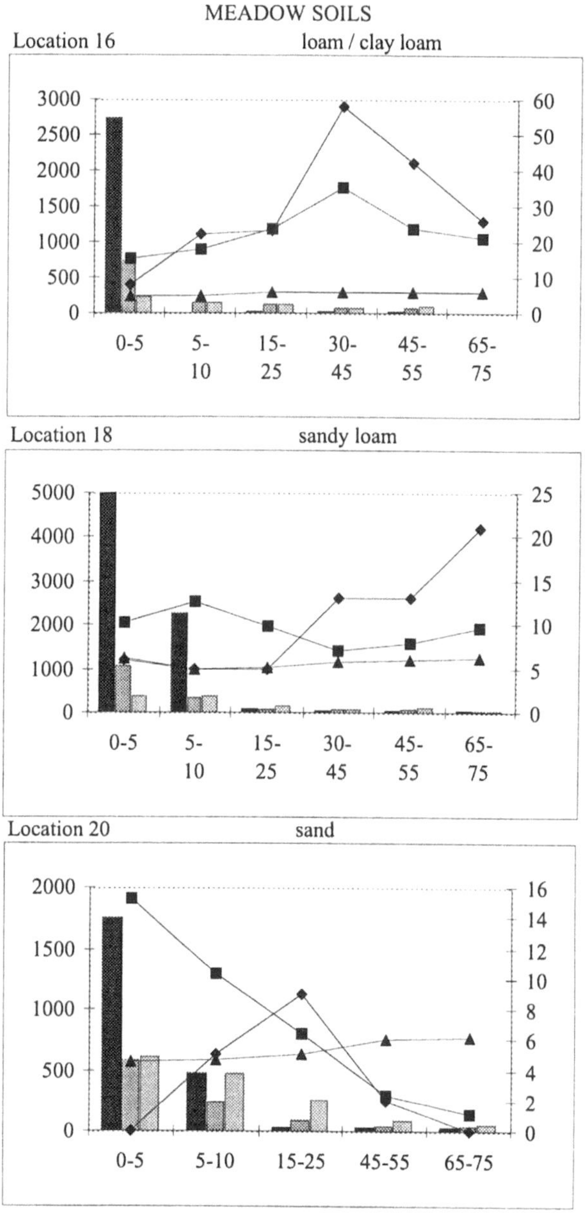
MEADOW SOILS
Location 16
loam / clay loam
3000
2500
2000
1500
1000
500
0
60
50
40
30
20
10
0
0-5
5-10
15-25
30-45
45-55
65-75
Location 18
sandy loam
5000
4000
3000
2000
1000
0
25
20
15
10
5
0
0-5
5-10
15-25
30-45
45-55
65-75
Location 20
sand
2000
1500
1000
500
0
16
14
12
10
8
6
4
2
0
0-5
5-10
15-25
45-55
65-75

6.6 Chemical Speciation of Metals in the Top-soil and Subsoil Layers

Numerous speciation procedures have been developed in order to describe heavy metal forms in soils (Hirner 1992; Rauret 1993; Zeien and Bruemmer 1989). In this study, the sequential extraction method according to Zeien and Bruemmer (1989, 1991) was applied to soil samples collected from sanitary protection zones. Both surface (0–20 cm) and subsurface (30–40 cm) soil samples were examined. The method involved the sequential extraction of 7 operationally defined fractions:

1. Soluble (mobile) – extracted by $1M$ NH_4NO_3 at indigenous pH
2. Easily mobilisable (exchangeable) – $1M$ NH_4OAc, pH 6.0
3. Occluded in MnO_x – $1\ M$ NH_2OH-HCl + $1M$ NH_4OAc, pH 6.0
4. Organically bound – $0.025\ M$ NH_4EDTA, pH 4.6
5. Occluded in amorphous FeO_x – $0.2\ M$ NH_4-oxalate, pH 3.25
6. Occluded in crystalline FeO_x – $0.2\ M$ NH_4-oxalate + $1M$ ascorbic acid, pH 3.25
7. Residual - aqua regia

Based on a sequence of extractants of increasing acidity, combined with reducing and complexing agents, this method was supposed to reflect quite well the chemical form of binding between metals and the soil substrate, and would also guarantee a minimum of redistribution of elements during extraction (Zeien and Bruemmer 1989, 1991; Karczewska et al. 1995).

Fraction 1 is likely to define the actual concentration of metals in soil solution. Fractions 2–4 might be mobilised in the short and medium term by changes in soil chemistry. Fraction 2 is supposed to be easily mobilised from the surface of soil constituents (soil sorptive complex) by changes in pH. Fractions 3 and 5, which represent metals bound to Mn oxides and amorphous Fe oxides, should be sensitive to changes in redox potential. Fraction 4, reflecting the organically bound metals, will probably be mobilised by the decomposition of soil organic matter. Fractions 6 and 7 are expected to be relatively stable, particularly in well-aerated soils, as reaction kinetics of iron oxides and silicates are slow.

The chemical analysis involved 5 metals of different origin: Cu, Pb, Zn, Fe and Mn. An attempt was made to examine Cd, the most hazardous metal emitted by the copper smelters. But, this was not successful, due to analytical problems (too low concentrations to be examined by flame AAS and very poor precision obtained by furnace graphite AAS).

Metal distribution among fractions 1–7 depended obviously on the metal itself, i.e. on its chemical properties, and on its origin in the soil (geogenic or anthropogenic, i.e. from metallurgical dust , see Tables 6.4, 6.5, and 6.6). Among the elements investigated, Fe and Mn showed considerable similarity throughout the whole spectrum of samples from different soil horizons, with a wide variety of soil properties. As expected, the predominating iron fractions in all samples were defined as in fractions 6, 5 and 7, which contributed altogether more than 94% to the total iron amount in the soil. Manganese turned out to be somewhat less uniformly distributed. In particular, the extremely poor sandy soils (sampling points 9 and 12) differed from other soils and showed relatively high content of soluble and exchangeable Mn forms (over 20%). In all other soils, apart from the sandy ones, the main fraction 3 accounted for more than 40% of soil manganese. Considerable amounts of Mn were also found in fraction 5. There was no essential difference between surface and subsurface samples with respect to manganese speciation.

Copper and lead, brought into soils mainly by the deposition of metallurgical dust, showed a much more diverse speciation, depending vitally on soil properties. However, it can be generally stated that the easily mobilisable fractions 1 and 2 contributed considerably to the total concentration of these metals in the soil. The soluble fraction 1 of copper constituted 2.2–3.7% of total Cu in all samples but sandy ones (and poor sandy loams). In the latter, soluble copper content was as high as 14.0–23.7%. The contribution of easily exchangeable Cu (fraction 2) varied in all samples from 14.8 to 44.2%, and was again highest in sandy soils. Despite the low concentrations of lead in the soluble fraction (fraction 1), which was 0.2–0.3% in most samples but sandy and sandy loams, the risk of lead mobilisation from polluted soils should be considered as relevant, because of the high proportion of exchangeable lead (fraction 2) , particularly in a surface soil layer (10.2–38.4%), and lower, but still significant, in a subsoil (2.3–13.2%). Although most of the lead in soils was bound in organic fractions (fraction 4) or occluded in manganese oxides and amorphous iron oxides (fractions 3 and 5), fractions 2 and 1 cannot be neglected, when considering the fate of lead in these soils.

High proportions of mobile Cu and Pb have been reported in polluted soils by others (Mc Bride 1989; Kociałkowski 1991; Zeien and Bruemmer 1991; Greinert 1993), suggesting that anthropogenic input of metals may contribute essentially to the total available metal pool, from where these metals may move downward in soil profile. However, this assumption seems to contradict the empirical observation for the stability and immobility of lead in soils .

Although concentrations in soils did not exceed the geochemical background, zinc turned out to be more mobile than copper. Mobile (fraction 1) and exchangeable (fraction 2) fractions of both metals corresponded very well with each other, reflecting first of all soil texture and the clay content. However, while fractions 4 and 2 were the main fractions of copper in most soils, the predominant concentrations of zinc were found in fractions 7 and 6, in sandy soils in fractions 6 and 2. Detailed results of metal speciation are presented in Tables 6.4–6.6 (on the following pages).

Table 6.4. Copper Speciation in some polluted soils differing in their texture and other properties

Location	Depth	Textural	Texture, %			pH	CEC	Corg	Cu total	Cu speciation – contributions of fractions to total Cu, %						
No.	cm	group	sand	silt	clay	1*M* KCl	mval/100g	%	mg/kg	1	2	3	4	5	6	7
9	0–20	sand	88	6	6	7.5	3.86	0.53	399.0	14.1	37.7	8.2	22.2	7.4	4.8	5.6
	30–40	sand	96	2	2	7.4	1.38	0.11	112.0	7.9	39.6	18.4	10.3	9.0	9.6	5.2
12	0–20	sand	93	5	2	7.9	2.86	0.37	426.0	14.0	38.4	10.3	18.2	9.3	4.7	5.1
	30–40	sand	97	2	1	7.8	1.00	0.05	7.4	0.0	44.2	28.4	2.1	3.1	20.2	2.0
13	0–20	sandy loam	63	33	4	5.1	4.81	1.31	889.0	23.7	34.1	11.7	12.9	10.1	2.1	5.4
	30–40	sand	91	6	3	5.8	2.05	0.24	not determined							
14	0–20	sandy loam	63	32	5	6.4	4.86	1.42	1710.0	24.4	33.3	7.5	12.8	8.2	3.3	7.4
	30–40	sand	92	5	3	6.4	2.37	0.26	35.0	5.7	26.0	11.4	27.1	10.0	14.2	5.7
7	0–20	silty loam	59	34	7	7.3	8.41	1.05	335.0	3.7	24.2	15.5	35.0	9.3	4.8	7.5
	30–40	sandy loam	83	13	4	6.8	4.55	0.34	65.0	2.2	14.8	11.7	44.0	12.6	7.7	7.0
4	0–20	silty loam	30	64	6	6.9	10.26	1.18	240.0	3.3	17.1	9.5	40.5	12.2	8.8	8.6
	30–40	silty loam	28	62	10	7.1	12.20	0.37	65.0	2.5	22.4	11.7	31.3	10.6	14.9	6.6
6	0–20	silty loam	42	48	10	5.8	9.36	1.10	558.0	2.5	20.5	15.7	29.5	15.9	7.2	8.7
	30–40	silty loam	40	51	9	6.2	8.32	0.52	not determined							
10	0–20	loam	33	39	28	6.7	42.2	2.58	271.0	2.3	23.6	15.6	43.2	11.1	6.9	7.3
	30–40	clay loam	28	38	34	6.5	46.3	1.57	49.5	2.5	19.6	16.0	32.8	16.8	5.1	7.2

Table 6.5. Lead and zinc speciation in some polluted soils differing in their texture and other properties

Location	Depth	Pb total	Pb speciation – contributions of fractions to total Pb, %							Zn total	Zn speciation – contributions of fractions to total Zn, %						
No.	cm	mg/kg	1	2	3	4	5	6	7	mg/kg	1	2	3	4	5	6	7
9	0–20	107.0	0.3	35.1	20.7	23.7	13.5	4.5	2.2	35.2	6.0	31.1	5.9	3.6	5.9	33.6	13.9
	30–40	11.2	0.4	8.6	14.9	32.0	28.7	12.9	2.5	7.3	2.7	2.7	5.4	0.0	17.5	23.5	48.2
12	0–20	149.0	0.5	38.4	24.3	23.0	9.8	1.2	2.8	40.8	7.0	32.8	7.5	4.3	17.2	8.6	22.6
	30–40	11.0	0.6	3.2	12.6	23.7	31.6	12.6	15.7	6.4	0.0	0.0	14.5	0.0	23.2	9.0	53.3
13	0–20	246.0	5.7	28.0	32.3	21.0	8.4	1.8	2.7	68.3	35.1	9.1	5.3	5.0	13.5	13.8	18.2
	30–40					not determined								not determined			
14	0–20	460.0	6.4	36.8	19.5	22.4	9.0	3.3	2.6	98.0	32.2	11.4	3.5	4.6	7.3	17.6	23.4
	30–40	85.0	0.5	12.9	17.6	24.1	27.9	11.8	5.2	36.0	0.6	9.7	7.6	5.4	11.1	25.4	40.2
7	0–20	114.0	0.2	11.1	29.3	37.3	16.2	2.8	3.1	48.0	5.0	13.9	10.3	6.0	5.1	27.7	32.0
	30–40	33.1	0.2	4.7	19.0	29.5	33.5	7.3	5.8	34.5	0.4	11.5	9.9	13.1	5.3	36.6	13.9
4	0–20	128.0	0.2	10.1	21.4	39.2	20.5	5.2	3.4	79.5	4.9	14.4	8.9	6.3	9.8	20.0	35.8
	30–40	42.0	0.2	2.3	24.0	33.4	27.3	10.0	2.8	50.5	6.8	6.6	5.5	4.2	9.5	24.1	43.3
6	0–20	233.0	0.2	16.4	35.3	33.7	8.7	2.5	3.1	153.0	4.3	16.5	9.7	8.0	19.2	22.4	19.8
	30–40					not determined								not determined			
10	0–20	122.0	0.2	18.5	26.0	37.9	13.1	3.1	2.2	92.4	3.8	11.8	8.8	6.6	12.3	25.1	32.6
	30–40	47.9	0.3	13.2	20.2	36.8	23.9	2.6	3.0	75.0	6.0	8.2	6.9	5.1	12.6	23.0	38.2

Table 6.6. Iron and Manganese speciation in some polluted soils differing in their texture and other properties

Location No.	Depth cm	Fe total mg/kg	Fe speciation – contributions of fractions to total Fe, %							Mn total mg/kg	Mn speciation – contributions of fractions to total Mn, %						
			1	2	3	4	5	6	7		1	2	3	4	5	6	7
9	0–20	3720	0.1	0.0	0.5	4.2	29.6	40.1	25.5	150.0	13.5	10.4	24.0	4.5	32.8	10.1	4.7
	30–40	2090	0.1	0.1	2.0	2.7	27.6	43.0	24.5	72.0	6.5	4.1	58.4	2.4	12.9	11.8	3.9
12	0–20	3430	0.0	0.0	0.9	4.0	28.3	33.1	33.7	130.0	13.6	8.9	23.5	3.7	32.6	9.2	8.5
	30–40	1710	0.3	0.2	2.9	3.0	25.9	40.1	27.6	55.0	5.8	3.2	63.8	2.5	8.6	9.7	6.4
13	0–20	5610	0.0	0.0	1.3	4.3	29.1	34.6	30.7	347.0	2.9	1.7	44.1	4.6	34.1	6.3	6.3
	30–40	not determined								not determined							
14	0–20	5330	0.0	0.0	1.2	6.1	23.0	39.6	30.1	296.0	4.0	5.1	30.1	5.6	38.4	9.4	7.4
	30–40	4820	0.1	0.1	1.2	2.5	27.2	42.0	26.9	186.0	3.3	4.9	49.9	2.0	20.2	11.5	8.2
7	0–20	8480	0.0	0.0	0.7	3.0	16.6	31.8	47.9	430.0	6.0	4.4	48.2	4.1	23.1	4.6	9.6
	30–40	8390	0.0	0.0	0.4	3.0	13.0	41.5	42.1	458.0	4.3	3.5	51.4	5.2	19.2	5.7	10.7
4	0–20	11240	0.0	0.0	0.3	2.6	12.6	40.6	43.9	402.0	6.3	7.4	45.0	3.5	17.4	8.1	12.3
	30–40	10420	0.0	0.0	0.3	1.9	8.6	44.6	44.6	370.0	5.4	2.2	54.1	3.7	13.6	8.9	12.1
6	0–20	17200	0.0	0.0	0.2	2.4	16.2	41.8	39.2	481.0	2.5	2.3	52.2	3.9	20.6	6.1	12.3
	30–40	not determined								not determined							
10	0–20	24200	0.0	0.0	0.2	2.2	18.2	43.6	35.8	730.0	1.8	2.0	53.8	3.6	18.5	6.5	13.8
	30–40	25600	0.0	0.0	0.2	1.9	12.1	45.9	39.9	645.0	1.9	2.6	56.2	3.0	17.5	6.9	11.9

6.7 Effect of pH and Redox Potential on Metal Speciation

Based on the results described above, and assuming that chemical sequential extraction could be used as an estimate for the binding form of metals in soils, it can be suggested that copper, lead and zinc may be sensitive in the examined soils to changes in soil chemistry. As the proportions of easily mobilisable metal fractions were relatively high in most of the samples, metal mobilisation should be considered as likely to happen even within a relatively short period of time. Such an effect could be caused for example by the changes in either soil reaction or redox potential. A decrease in soil pH will result in enhanced metal mobility, as it is widely known and documented in numerous publications (Alloway 1986; Gorlach and Gambuś 1991ab; McBride 1989; Kabata-Pendias and Pendias 1993). Some studies also addressed this relationship with respect to the metal-contaminated soils surrounding the copper smelters in LGOM (Kabata-Pendias 1979; Komisarek 1990; Greinert 1993). All authors emphasized the strong dependence of metal mobility on soil reaction. According to this, the contaminated soils of the sanitary protection zones have been limed in order to prevent a possible drop in soil pH.

Redox potential, which should be considered as another factor governing changes in soil chemistry and the transformation of solid metal associations, has not been equally examined yet. Only a few publications have addressed this issue (Gliński et al. 1986, 1992; Jackowska and Gąszczyk 1990; Kukier et al. 1990) by examining unpolluted or slightly polluted soils. However, this problem seems to be of significant environmental concern, especially with regard to strongly polluted soils. Oxygen depletion in flooded soils should cause a successive reduction of inorganic soil compounds, i.e. manganese oxides at first and then iron oxides, thereby releasing other heavy metals occluded to these oxides (Barlett 1986; Gliński et al. 1992; Kabata-Pendias and Pendias 1993; Rauret 1993).

In order to assess metal mobility and possible changes in the chemical form of metals as a result of a decrease in redox potential, a pot experiment is now being carried out, dealing with polluted soils from the neighbourhood of the copper smelter "Głogów". The experiment involves two strongly polluted soils differing in clay content, with three different ranges of soil reaction, as well as various watering regimes, i.e. mild, wet and extremely wet. In the latter, a significant drop in soil Eh was expected. Since the experiment started last year and was designed for a two-year period, no final results can be presented yet. However, based on concentrations in the soil solution as well as on the uptake of metals by wheat, grown in the pots (see: Table 6.7), changes of the main forms of heavy metals as well as the intensity of metal mobilisation tend to be both dependent on soil pH and the watering regime. Final results of this experiment will be published elsewhere.

Table 6.7. Mobility and bioavailability of heavy metals under different pH and moisture conditions (preliminary results of the pot experiment)

Soil No.	Soil pH	Watering conditions	Metals in soil solution, mg/dm^3								Metals in plant material, mg/kg dw.							
			After 30 days				After 90 days				Wheat straw				Wheat grains			
	1*M* KCl		Cu	Pb	Zn	Cd	Cu	Pb	Zn	Cd	Cu	Pb	Zn	Cd	Cu	Pb	Zn	Cd
4	5.1	wet	0.22	0.17	0.30	0.02	0.35	0.17	0.21	0.03	11.40	3.20	180	0.80	12.00	0.40	92	0.60
Legnica	4.8	mild	[a]	[a]	[a]	[a]	0.38	0.15	0.33	0.05	12.80	4.40	189	0.80	12.10	0.70	105	0.70
	7	wet	0.19	0.11	0.17	0.01	0.29	0.07	0.06	0.01	9.00	3.00	132	0.80	11.90	0.40	100	0.50
	6.7	mild	[a]	[a]	[a]	[a]	0.31	0.16	0.15	0.02	12.00	3.40	140	0.80	12.50	0.60	104	0.60
	7.5	wet	0.12	0.12	0.02	[b]	0.39	0.22	0.10	[b]	8.80	2.20	185	0.80	12.50	0.20	96	0.20
	7.5	mild	0.12	[a]	0.06	[b]	0.30	0.19	0.18	[b]	10.00	2.60	187	0.80	12.10	0.60	100	0.40
13	5.3	wet	1.05	0.14	4.00	0.03	5.80	0.22	14.60	0.06	plant material unsuitable for analysis				wheat did not bear grains			
Głogów	4.6	mild	[a]	[a]	[a]	[a]	10.00	0.31	18.30	0.12								
	6.6	wet	0.73	0.40	2.80	0.03	[a]	0.17	17.90	0.08								
	6.4	mild	[a]	[a]	[a]	[a]	[a]	0.22	14.90	0.10								
	7.4	wet	0.25	0.17	0.85	0.01	0.30	0.12	2.13	0.03	60.00	6.20	70	1.60				
	7.3	mild	[a]	[a]	[a]	[a]	0.60	0.16	3.57	0.03	75.00	11.80	106	2.20				

[a] soil solution not obtained
[b] below detection limit (i.e. < 0.01 mg/dm^3)

The pot experiment was preceded by a simplified laboratory approach. For this, changes in metal forms were determined after a 20-day incubation of soils under flooded conditions. Two soils of different textures and other properties, and consequently unlike in their initial metal speciation pattern, were taken from sampling points 4 and 13 (Fig. 6.1). In 100-cm^3 beakers, 100 g of air-dried samples were placed and flooded in two different schemes, i.e. with distilled water (scheme B) and with 0.5% glucose solution (scheme C), as the intensity of anoxic microbiological processes was believed to depend on the content of easily bioavailable organic compounds (Barlett 1986; Gliński et al. 1992). All samples were incubated in air-tight sealed beakers at room temperature and for a period of 20 days. Redox potential and pH were measured continuously after incubation and then all samples were centrifuged and air dried. Metal concentration was determined in supernatants and all soil samples were sequentially extracted according to Zeien and Bruemmer. The extraction results are presented in Table 6.8 and Fig. 6.4. Scheme A refers to untreated soil samples. In all schemes, the redox potential decreased during incubation, especially in the samples which were flooded with glucose solution. Soil pH increased by 0.6–1.0 unit due to flooding with water. In the samples with glucose, the final pH was much lower and in the case of soil no. 4, even lower than the initial soil pH, reflecting the complexity of factors governing the changes of soil properties under flooded conditions. All these effects were significant at the $p=0.01$ level.

Relevant changes in metal forms were observed in the experiment. Anaerobic incubation resulted in dissolution of certain metal amounts from soil. Released amounts of iron and manganese were much higher than the "easily soluble" fraction 1 of these metals determined in soil samples, which confirmed the effects of reduction processes. Unlike iron and manganese, the released amounts of other metals i.e. copper, lead and zinc, were lower than their "easily soluble" fraction 1, though undoubtedly affected by soil–soil solution equilibria. In all cases, much higher amounts of metals were released with glucose, which must have been caused both by lower pH and Eh values of the system and probably also by a more intensive microbiological activity. The changes in metal speciation, observed in scheme C, were also much stronger than in B. Changes in metal speciation concerned fractions 1–5 as it is shown in Table 6.8. Apart from releasing the most mobile metal fraction from soil, also the contributions of fractions 1 and 2 together, in the non-released metals increased in soil no 4 up to: 195 mg/kg Cu, 88 mg/kg Pb and 45 mg/kg Zn, and in soil no 13 up to 488 mg/kg Cu and 117 mg/kg Pb. As expected, the contributions of fraction 3 decreased significantly after incubation. Such a transformation seems to be relevant if considering the risk of metal mobilisation from soil into underground water. Potential toxic effects of released metals on plants, including manganese and iron should not be neglected either, but this problem exceeds the topic of this chapter.

Table 6.8. Changes in heavy metal speciation in polluted soils as an effect of simulated flooding (preliminary results of the laboratory experiment)

Soil	Scheme	Metal	Total metal content mg/kg	Speciation - contributions of fractions, mg/mg 1	2	3	4	5	6	7
4	Control	Cu	558	13.9	114.3	87.6	164.6	88.7	41.9	48.5
Legnica	A*	Pb	233	0.6	38.3	82.2	78.6	20.2	5.8	7.3
		Zn	153	6.6	25.3	14.8	12.3	29.4	34.3	30.3
		Fe	17200	0.2	0.6	33.8	416	2780	7190	6740
		Mn	481	12.2	11.1	251	18.7	99.3	29.3	59.4
	Flooded	Cu	526	11.5	108.8	51	158.3	95.7	52.1	48.3
	with	Pb	230	3.7	47.9	64.7	77.4	22.2	6.1	8
	water	Zn	150	6.2	25.1	12	12.3	31.2	33.9	29.3
	B*	Fe	17100	0.2	1.9	21.8	435	2680	7340	6620
		Mn	475	77.3	29.7	168	13.2	98.4	30.6	57.8
	Flooded	Cu	503	75.3	119.9	42.6	106.2	76.9	36.2	45.9
	with	Pb	230	23.6	64.3	42.9	62.4	23.3	5.7	7.8
	glucose	Zn	141	38.2	6.5	2.3	4.6	22	33.8	31.6
	solution	Fe	16580	30.9	4.7	97.6	447	2080	6980	6830
	C*	Mn	356	198	9	19.5	3.1	33.2	30.6	62.8
13	Control	Cu	889	209.2	303	107	113.5	89.2	19.3	47.8
Głogów	A*	Pb	246	14.1	68.9	79.6	51.6	20.6	4.5	6.7
		Zn	68.3	24	6.2	3.6	3.4	9.2	9.4	12.4
		Fe	5610	0.7	2.6	71.3	243	1630	1940	1720
		Mn	347	10.2	5.8	153.2	15.8	118.4	21.9	21.7
	Flooded	Cu	849	188.5	292	92.3	114.6	96.3	18.7	46.6
	with	Pb	242	12.7	69.2	76.9	50.3	22.2	4.3	6.4
	water	Zn	66.4	24.6	7.3	2.4	3.9	8.8	8.2	11.2
	B*	Fe	5400	1	3.2	66.2	272	1390	1980	1680
		Mn	331	24.8	2.2	130.2	16.6	106.3	20.8	10.5
	Flooded	Cu	810	235.1	242.8	54.7	96.3	105.2	21.6	54.3
	with	Pb	236	25.4	86.2	36.4	54.2	22.6	4.6	6.6
	glucose	Zn	62.2	22.3	5.6	1.2	3.7	9.2	8.6	11.6
	solution	Fe	5310	1.8	5.5	75.4	356	1260	2050	1560
	C*	Mn	256	145.5	6.8	3.2	3.3	56.2	20.3	20.7

* Final pH and Eh values in schemes:

	A	B	C	A	B	C
	Sample 4			Sample 13		
pH:	6.51	7.09	5.34	5.25	6.22	5.77
Eh. mV:		300	260		390	220

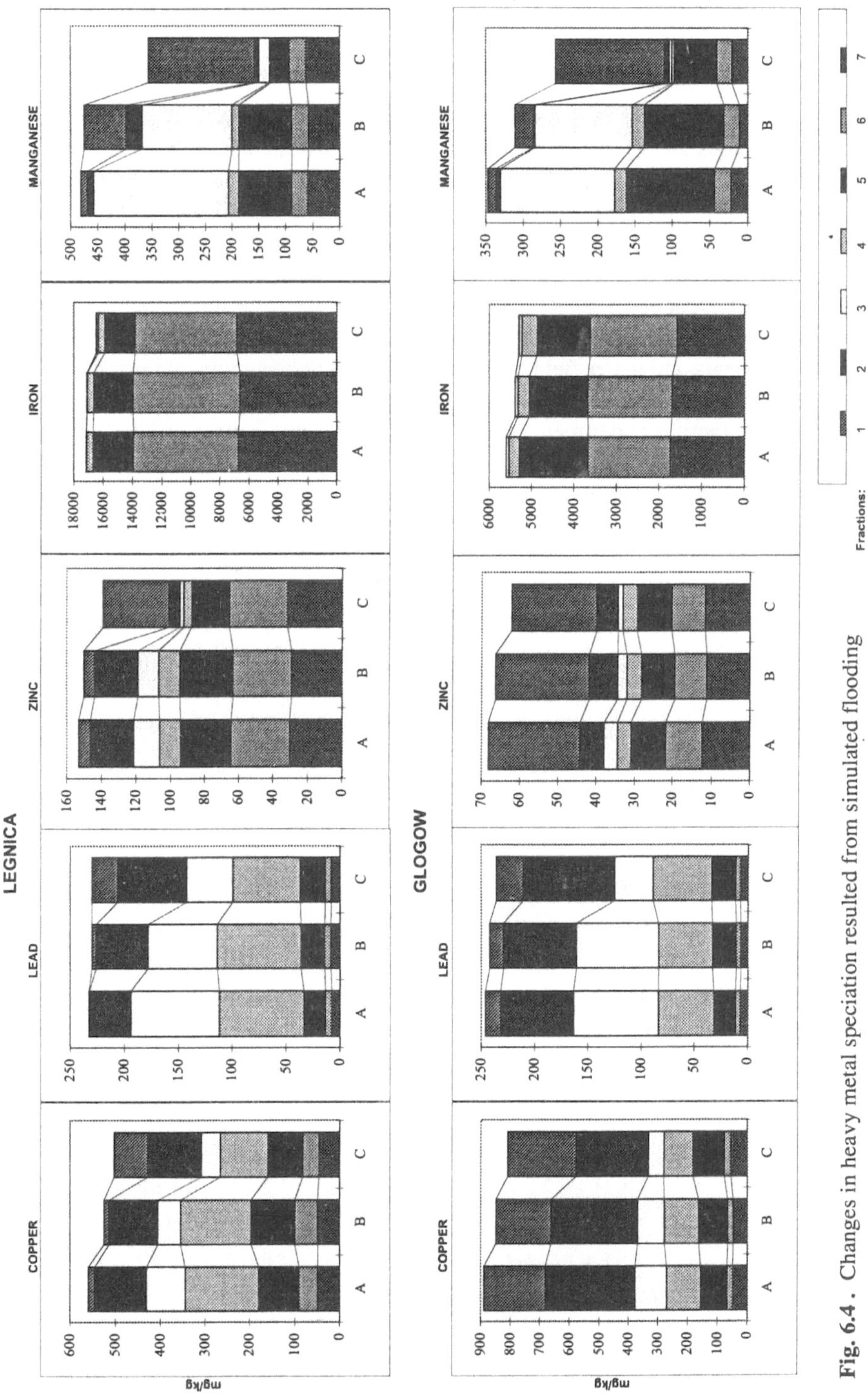

Fig. 6.4. Changes in heavy metal speciation resulted from simulated flooding

6.8 Other Approaches to Metal Forms and Mobility

Finally, some data on heavy metal concentration in plants grown within the sanitary protections zones will be presented, as they illustrate the problem of current environmental pollution in the surroundings of copper smelters. However, it should be emphasized that not only the metal uptake from soil, but first of all direct dust deposition contribute to final metal concentration in the plant tissues. Table 6.9 presents data on Cu, Pb, Zn and Cd concentrations in the straw and grains of different cereals collected in years 1989–93 from the "fixed sampling points" described above. Although chemical compositions of plants grown in the pot experiments and under field conditions should not be simply compared, it can be easily seen that Cu and Pb concentrations in the straw were much higher in the field that in the pot experiment, where soil was the only source of heavy metals for plants.

In the end, it is worth mentioning that another kind of approach could still be undertaken to determine copper and lead forms in soils polluted with metallurgical dust. This is namely the direct electron microprobe analysis (Hiller and Bruemmer 1989; Weber and Kowaliński 1987; Weber 1989). However, some attempts to apply this method to the soils from copper smelter neighbour-

Table 6.9. Heavy metals concentration in cereal samples in the fixed points situated in the sanitary protection zones (1989–93)

Smelter	Distance, km Direction	Location No.	Metals in cereals, mg/kg d.w. Straw Cu	Pb	Zn	Cd *	Grains Cu	Pb	Zn	Cd *
Legnica	0.5 SE	1	7.0–21.3	4.6–14.4	18.3–34.8	0.7	0.6–6.7	1.3–3.6	1.3–12.0	0.52
	1.0 W	2	46.2–60.0	7.2–14.5	92–120	–	7.9–11.5	3.0–12.9	22.3–58.7	–
	2.0 W	3	3.3–19.8	7.1–8.0	7.3–47.6	4.7	0.3–10.0	2.3–7.2	6.2–34.5	0.65
	3.0 W	4	4.8–14.2	2.5–10.0	7.1–44.0	1.22	0.4–9.0	0.1–4.0	13.2–46.5	0.61
	4.0 W	5	3.1–15.4	5.3–8.0	6.7–29.2	0.73	0.4–8.0	0.8–4.3	14.0–42.0	0.31
	0.5 NW	6	no data				no data			
Głogów	1.0 SE	7	no data				no data			
	2.0 SW	8	2.9–14.4	2.1–8.3	8.1–17.3	0.84	1.2–3.8	0.4–6.6	16.8–32.0	0.46
	3.0 NE	9	4.1	3.2	27.3	1.73	0.8	2.2	19.1	0.68
	4.0 NE	10	4.1–6.4	1.9–9.6	8.7–12.5	2.03	0.6–5.0	0.7–5.1	10.1–34.0	0.62
	4.0 SE	11	3.2–5.0	4.0–6.2	15.0–31.2	1.12	1.1–5.2	1.3–2.2	20.8–35.8	0.32
	0.5 SE	13	no data				no data			

* one analysis only

hoods did not give any satisfactory quantitative information, as the analysis left out an account of the most dispersed metal fraction, which is the most important one from an ecological point of view (Weber and Kowaliński 1987).

6.9 Conclusions

- A strong accumulation of heavy metals, especially Cu and Pb, occurred in the surface layer of soils surrounding the copper smelters "Legnica" and "Głogów".
- Although no indication was found in the profile distribution of heavy metals in soils which would support the occurrence of metal leaching, it is very likely that, in spite of soil liming, such a process takes place, especially in sandy soils.
- A high contribution of easily mobilisable fractions of Cu, Pb and Zn in the examined soils confirmed the enhanced risk of metal mobilisation.
- Changes in soil redox properties, in particular due to soil flooding, may result in the dissolution and mobilisation of heavy metals from the polluted soils.

References

Adriano DC (1986) Trace elements in the terrestrial environment. Springer, Berlin Heidelberg New York

Alloway BJ (ed)(1986) Heavy metals in soils. Blackie, Glasgow

Bartlett RJ (1986) Soil redox behaviour. In: Sparks DJ (ed) Soil physical chemistry. CRC Press, Boca Raton Fl. 179–207

Borkowski J, Mazur B, Szerszeń L (1991) Proceedings of symposium: geological aspects of environmental protection, Kraków 21–23.10.91, AGH Kraków, 77–82 (in Polish)

Czarnowska K, Gworek B (1991) The consistent presentation of geochemical background of trace elements in the arable soils developed from sedimentary rocks. In: Proceedings of 7th symposium on microelements in agriculture, held in Wrocław 9–10.9.1989, Wyd AR Wrocław, 37–41 (in Polish)

Dudka S. (1992) Factor analysis of total element concentrations in surface soils of Poland. in: The science of the total environment, Elsevier Science Publ.B.V., Amsterdam, 121:39–52

Gliński J, Benicelli R, Stępniewska Z (1986) Changes in the oxygen conditions of the soil of different degrees of compaction subjected to water and drying under the conditions of a model experiment. Polish J Soil Sci XIX 21–26

Gliński J, Stahr K, Stępniewska Z, Brzezińska Z (1992) Changes of redox and pH conditions in a flooded soil amended with glucose and nitrate under laboratory conditions. Z Pflanzenernahr Bodenk 155:13–15

Gorlach E, Gambuś F (1991a) Desorption and phytotoxicity of heavy metals depending on soil properties (in Polish). Rocz Glebozn 42:207–214

Gorlach E, Gambuś F (1991b) The effect of liming, adding peat and phosphorous fertilization on the uptake of heavy metals by plants. Pol J Soil Sci 24/2:199–204

Greinert H (1993) The solubility and plant availability of heavy metals in soils contaminated with industrial dusts and sludges. In: Eijsackers, Hamers (eds) Integrated soil and sediment research. A basis for proper protection. Kluwer, Dordrecht, 257–260

Hiller DA, Bruemmer GW (1989) Electron microbeam analysis for the determination of heavy metal bound forms in soils of varying heavy metal pollution. Mitt Dtsch Bodenkundl Gesellsch 59:1179–1182

Hirner AV (1992) Trace element speciation in soils and sediments using sequential chemical extraction methods. Intern J Environ Anal Chem , 46:77–85

Hornburg V, Bruemmer G.W (1993) Verhalten von Schwermetallen in Boeden:
1. Untersuchungen zur Schwermetallmobilitaet. Z Pflanzenernaehr Bodenk, 156:467–477

Jackowska I, Gąszczyk R (1990) Desorption of Fe^{2+}, Fe^{3+} and Mn^{2+} cations under differentiated oxygen and pH conditions in the brown soil developed from heavy loam. Zesz Probl Post Nauk Roln 388:261–267

Kabata-Pendias A (1979) Effect of lime and peat on heavy metal uptake by plants from soils contaminated by an emission of copper smelter. Rocz Glebozn, 30/3:323–328

Kabata-Pendias A, Bolibrzuch E, Tarłowski P (1981a) Impact of a copper smelter on agricultural environment. Part 1: Contamination of soils. Rocz Glebozn 32/3:207–214 (in Polish)

Kabata-Pendias A, Bolibrzuch E (1981b) Impact of a copper smelter on agricultural environment. Part 3: Contamination of cultivated plants. Rocz Glebozn 32/3:223–228 (in Polish)

Kabata-Pendias A, Pendias H (1993) Biogeochemia pierwiastków śladowych. PWN, Warszawa (in Polish)

Karczewska A (1995) Metal species distribution in the area affected by copper smelter. Appl Geochem (submitted)

Karczewska A, Wenzel WW, Mavrodieva R (1994) Effect of metal sources and indigenous soil pH on metal fractions in soil. Environ Geochem Health, (submitted)

Kociałkowski WZ, Komisarek J, Grzebisz W (1991) Methods of heavy metal fractionation in soils polluted by copper metallurgy. Proceedings of symposium: Geological aspects of environmental protection, Kraków 21–23.10.91, AGH Kraków, 135–139 (in Polish)

Komisarek W (1990) Effect of $CaCO_3$ on the content of various forms of Cu, Zn and Pb in contaminated soils. Wyd Pozn TPN, Nauki Roln, 69:53–62 (in Polish)

Kowaliński S, Drozd J, Licznar M (1979) The influence of the distance of a copper smelter on the physico-chemical properties of soils. Pol J of Soil Sci, 12:1–15

Kowaliński S, Bogda A, Borkowski J, Chodak T, Drozd J, Licznar M, Roszyk E (1972) Research on the effects of industrial emissions of copper smelter "Legnica" on the changes of some soil properties. Proceedings of 19th Congress of Polish Soil Science Society (PTG), PTG Warszawa (in Polish) 296–318

Kukier U, Turski R, Kasiak A (1990) Model investigations on the iron reduction. Rocz Glebozn 41:59–66

Mc Bride M.B. (1989) Reaction controlling heavy metal solubility in soils. In: Advances in soil science. Springer, Berlin Heidelberg New York, 10:1–56

Rachwal L (1990) The fractionation of copper, lead and zinc and their distribution in various depths of soils in the protection zones of copper smelters in Głogów. Wyd Pozn TPN, Nauki Roln, 69:101–114 (in Polish)

Rauret G (ed)(1993) Workshop on sequential extraction in sediments and soils. Selected papers. Intern J Environ Anal Chem., (special issue), 51

Roszyk E, Szerszeń L (1988a) Accumulation of heavy metals in the arable layer of soils of the sanitary protection zone in the vicinity of copper metallurgic plants. Part 1: Legnica. Rocz Glebozn 39,4:135–141

Roszyk E, Szerszeń L (1988b) Accumulation of heavy metals in the arable layer of soils of the sanitary protection zone in the vicinity of copper metallurgic plants. Part 2: Głogów. Rocz Glebozn 39,4:147–156

Shamsham S (1992) Distribution of some metals in arable soil profiles in the region contaminated by copper industry. Part 1. Soil characteristics. Rocz Glebozn 43, 3–4: 125–131

Shamsham S (1993) Distribution of some metals in arable soil profiles in the region contaminated by copper industry. Part 2. Total content of metals. Rocz Glebozn 43, 1–2: 65–71

Szerszeń L, Chodak T, Karczewska A (1993) Areal, profile and time differentiation of heavy metal content in soils in the vicinity of copper smelters in LGOM, Poland. In: Eijsackers, Hamers (eds) Integrated soil and sediment research. A basis for proper protection. Kluwer, Dordrecht, 279–281

Szerszeń L, Karczewska A, Roszyk E, Chodak T (1991) Distribution of Cu, Pb and Zn in profiles of soils adjoining copper metallurgic plants (in Polish).RoczGlebozn., 42, 3/4:199–206

Weber J (1987) Ecological condition of soils in Lower Silesia. In: Ecological state of Lower Silesia. Proceedings of symposium, Wrocław 1987, DTSK Wrocław

Weber J, Kowaliński S (1987) Suitability of scanning electron microscope and electron microprobe analyser to heavy metal polluted soils investigations (in Polish). Arch Ochr Œr 1–2, 193–206

Weber J (1989) Chemical and micromorphological properties of soils under the influence of copper industry. Proceedings of International Conference on Soil Conservation and Environment, Piestany-Spa, Slovakia, 8–11

Zeien H, Bruemmer GW (1989) Chemische Extraktionen zur Bestimmung von Schwermetallbindungsformen in Boeden. Mitt Dtsch Bodenkundl Gesellsch, 59/i:505–510

Zeien H, Bruemmer GW (1991) Ermittlung der Mobilitaet und Bindungsformen von Schwermetallen in Boeden mittels Sequentieller Extraktionen. Mitt DtschBodenkundl Gesellsch, 66/I: 439–442

7 Heavy Metal Contamination of Soils: Reuse versus Disposal

J. Matschullat

7.1 Introduction

More and more, sites of heavily contaminated soils are being detected in most industrialized countries. In Germany alone, about 138 722 sites were registered within the past few years, while estimates of 244 926 sites are being discussed (data from Umweltbundesamt, Berlin 1994). According to the increased standards for polluted soils (e.g. German sewage sludge standards/Klärschlammverordnung from 1992), demand is high to find solutions to the problem of safe contaminant disposal.

The question should be discussed, however, whether all of the contaminated soils from these sites should be treated as waste material. The demand for safe repository space, incinerator or cleaning capacity etc., far exceeds the existing capabilities. This is but one reason to further discuss all remedial options. Another argument is related to land use. While some sites may have to be used for agricultural production or private housing, others may serve as a future shopping mall or a new industrial site with sealed surface areas. Different uses require differentiated answers, which will have to consider more factors than threshold values of soil contaminant levels.

Nevertheless, standards for environmental quality are a useful and necessary tool to assess an unknown site and to briefly characterize its potential hazard. If these standards were always taken literally, however, large regions like some former mining areas would have to be evacuated immediately. But nobody really thinks of abandoning for example, the small German towns in the western Harz mountains, where most soils largely exceed the sewage sludge limits (see Fig. 7.1; Nowak and Preul 1971). Do officials there neglect their duties?

A land use related example of a contaminated site shall highlight the necessity of site-specific decisions versus the global application of standard threshold values.

7.2 A Lead Processing Factory – The Culprit?

A lead processing factory in the Harz mountains, Germany, intends to enhance its productivity and build an extension to existing halls on their property (Fig. 7.1). Soil investigations reveal a moderate to high contamination of As, Ba, Cu, Pb, Sb, and Zn (see Tables 7.1 and 7.3). The site is situated on a river bank (lower and middle terrace) and protected from flooding. Percolation water drains the site in the direction of the adjacent stream.

Table 7.1. Concentrations in the investigated soil samples (n = 42) in comparison to the probable local geochemical background (BG) and current soil contamination levels (SCL)*

Element	Sample [μg/g]	Median	BG [μg/g]	SCL [μg/g][a]
As	7 - 970	130	50	50/-
Ba	440 - 16 700	1200	350	2.000/-
Cu	<5 - 900	270	110	500/800
Pb	20 - 23 700	1270	50	600/900
Sb	<5 - 240	50	no data	no data
Se	<10	<10	no data	no data
Zn	40 - 1800	390	230	3000/2500

[a] The values (x/) are taken from the "Leitraad Bodensaneering 04.11.1988 of the Netherlands, category C": threshold values where a cleanup is considered necessary; while the second set of values (/x) refers to the maximum concentrations in sewage sludge according to the German sewage sludge standards from 1992. The background values are taken from (Matschullat et al. 1991) and unpublished data from the author

While the median concentrations for Pb exceed the German threshold values, all other element concentrations are either not represented by the German standards (As, Ba, Sb, Se) or can be neglected in respect to these standards (Cu, Zn). Nevertheless, exceeding concentrations of one element alone are sufficient to require a cleanup of the respective site. A costly enterprise, which triggers the question as to the culprit. Hypothetically, three different pollution sources can be identified for the present soil contamination:

1. Remains of the historical mining activities in and around the city of St. Andreasberg (Ag-, Pb-, Zn-ores),
2. Remains of an ammunition factory from the 1930s which was situated on the site, and
3. Waste materials released since the 1940s by the factory itself.

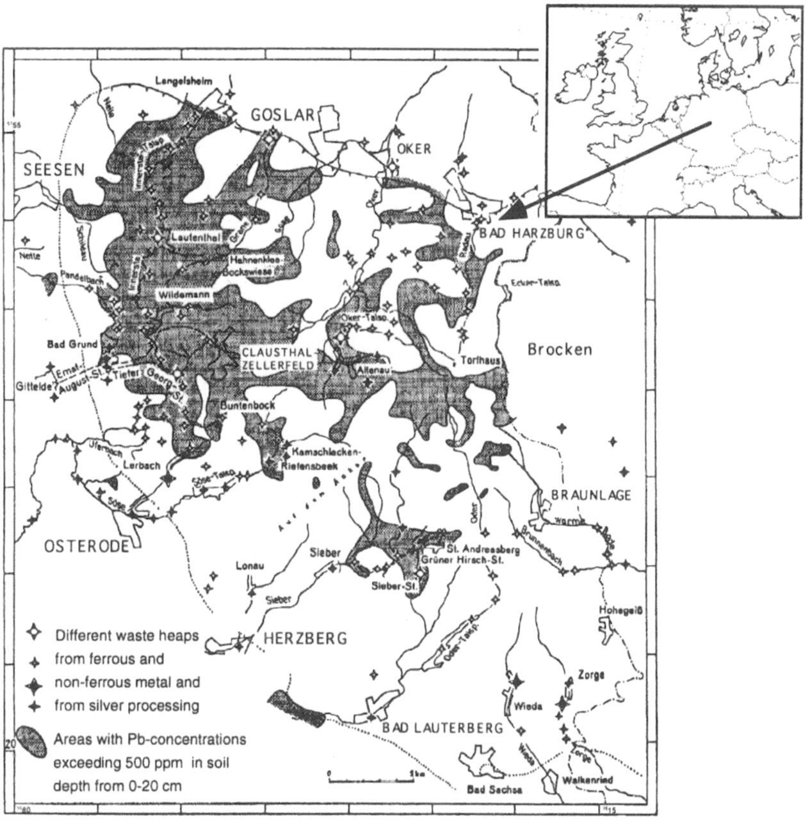

Fig. 7.1. Position of the factory (**X**) in the Harz mountains, relative to the mining area of St. Andreasberg. The larger figure is from Nowak and Preul (1971)

1. Mining: For centuries, mining of hydrothermal ores was the economic spine of St. Andreasberg, a small town several kilometers upstream from the factory. Lead- and zinc ores with higher concentrations of silver and copper were exploited. Accompanying minerals yield antimony and arsenic, while barium is the dominant matrix ore. Exact data on the chemical composition of the ore could not be found. Because of geological similarities with the ore province of the neighbouring areas of Clausthal-Zellerfeld and Lautenthal, the following data seem valuable: The Pb/Zn ratio is 0.1 to 20, the Pb/Cu ratio 4 to 8. No data could be found for Sb, Se, and Ba (Table 7.4).

2. Ammunition plant: There is no information available about the possible element dispersion and waste materials from such a plant. It is unlikely, however, that this factory will have contributed to the geochemical fingerprint of the soils. A screening for organic contaminants did not show anomalies.

3. Lead processing factory: As, Ba, Cd, Cr, Hg, Mn, Ni, Pb, Sb, and Se play a vital role in the production of modern lead-based materials (Hiller et al. 1990; Kinner et al. 1986). In this case, only As, Ba, Pb, Sb, and Se have to be considered. According to data from the producer, Sb is present with 5–11.5 weight-%, and As with 0.02–0.2 weight-%, Ba-sulphate with up to 0.15 weight-%, Bi with 0.03 weight-%, and Se with 0.01–0.025 weight-%. Following factory data, the lead parts are composed of 90–95% Pb, 5–10% Sb, 0.05–0.3% As, and 0.05% Cu. In addition, 0.29% Ba-sulphate is being used (see Table 7.2).

Table 7.2. Concentrations (weight-%) of the investigated elements present in lead processing. Data courtesy of the factory management

Element	Average lead processing	This factory
As	0.02 - 0.2%	0.05 - 0.3%
Ba	≤ 0.15%	0.29%
Cu	no data	0.05%
Sb	5 - 11.5%	5 - 10%
Se	no data	0.01 - 0.025%
Pb	90 - 95%	90 - 95%

These results show that the elements As, Ba, Cu, Sb, Pb, and Zn are relevant for the hypotheses. No Selenium enrichment could be detected (see Table 7.1). It is important to note the large concentration range of the other six elements over two to four orders of magnitude (see Table 7.1). This inhomogeneity can be observed both within a soil profile as between several excavations. Neither vertical nor horizontal profiles revealed trends in element enrichment (Table 7.3, Fig. 7.2).

7.3 Materials and Methods

All samples were taken with plastic shovels from fresh soil excavations (maximum depth 200 cm) and stored in polyethylene cups. The samples were oven-dried (60 °C) to constant weight and subsequently sieved (<63 μm/240 mesh) with nylon sieves. This fine fraction was used for the geochemical analysis after making compressed powder-discs for X-ray fluorescence spectrometry. The more coarse material consisted of non-ore mineral bearing rock and river sediment material. The fraction (<63 μm) constituted an average (median) of 5% of the total sample weight (range: 2–20%). The samples were measured with the spectrometer PW 1480 (Philips) and the signals processed using the multi-element analytical system UniQuant® (De Jongh). Three international standard reference materials and multiple sample preparation were used for quality control (the standards GXR 2 and GXR 6 from soils; the standard GXR 1, a jasperoid mineral). The results are satisfactory, except for copper in concentrations <50 μg/g (Govindaraju 1989).

7.4 Results and Discussion

Usually, the geogenic background for an unknown soil is indicated by the chemical composition of a "typical, average soil", e.g. international data or even crustal abundance (Bowen 1979; Wedepohl 1995). These data are unrealistic for the western Harz region. Localities in the Harz require the use of local or regional

Table 7.3. Samples from the excavations (depth given in cm; concentrations in μg/g).

Sample	Depth	As	Ba	Cu	Pb	Sb	Zn
B1-1	30	7	520	<5	20	<5	40
2	60	12	660	<5	80	<5	85
3	90	13	720	<5	90	<5	40
4	120	165	13.500	390	19.000	130	740
B2-1	30	450	2.800	380	5.300	120	710
2	80	310	2.900	270	6.000	80	590
3	110	10	550	<5	50	<5	40
4	160	10	770	<5	220	3	90
5	200	640	1.500	810	4.500	210	900
B6-1	20	90	990	34	16.100	60	300
2	40	940	2.600	650	7.000	140	1.800
3	100	490	2.400	520	4.300	130	1.200
4	150	10	770	<5	220	3	90
5	200	520	1.600	700	3.600	190	1.200
B15-1	20	20	610	<5	100	<5	70
2	60	970	1.500	900	5.300	240	1.300
3	100	130	1.200	340	1.100	50	880
4	150	20	1.100	<5	90	<5	380
5	180	40	930	<5	100	10	380

background values which are influenced by the Palaeozoic geology including Late Palaeozoic and Mesozoic mineralizations. In addition, the long period of active mining (throughout the past 2 millennia) altered the background considerably by diffuse element dispersion. More realistic data – rounded median values – for this region are: 50 μg As/g (130), 350 μg Ba/g (430), 110 μg Cu/g (180), 50 μg Pb/g (74), and 230 μg Zn/g (280) (see Table 7.1; Matschullat et al. 1991). Data in parenthesis refer to maximum background concentrations.

Additional, unpublished investigations by other workers were focused on lead and zinc concentrations - and partly cadmium - within the factory site and in its vicinity several kilometers up- and downstream. They show high concentrations for these two (three) elements that can directly be related to mining activities. Cadmium concentrations in this work show median values of 10 μg/g. There will be no further discussion on this element due to the comparatively low concentrations in respect to the detection limits of the XRF-method used.

Single samples show high concentrations of the elements As, Ba, Cu, Pb, Sb, and Zn (see Table 7.3). Some of these concentrations might be influenced by emissions from the recent lead processing activity. This cannot be verified, however, as the possible ranges from natural dispersion and migration processes of the relevant elements allow similar correlations. No sample shows anomalies, which doubtlessly correlate with the lead processing of this factory (Tables 7.2 and 7.4).

Table 7.4. Ratio values for lead processing, for the investigated soil samples (median values given in parenthesis), and for average soils and hercynian ores.

	Lead processing	Soil samples		Average soil	Local background[a]	Hercynian ore
Pb/As	300–1900	2–180	(7)	6	1	no data
Pb/Ba	200	0.1– 16	(1)	0.1	0.1	no data
Pb/Cu	1800–1900	3–470	(20)	1	0.5	4–8
Pb/Sb	10–20	10–270	(50)	35	no data	no data
Pb/Zn	no Zn	1– 30	(3)	0.4	0.2	0.1–20

[a] for local background concentrations see Table 7.1.

Comparing the element ratios of lead processing with the element ratios of (1) the investigated soils, (2) an average soil, and (3) the hercynian mineral ores, the investigated soil shows a closer relationship with the average soil and the hercynian ores (see Table 7.4). The data support the hypothesis of the historical contamination of the site from mining activities upstream and the subsequent element dissipation.

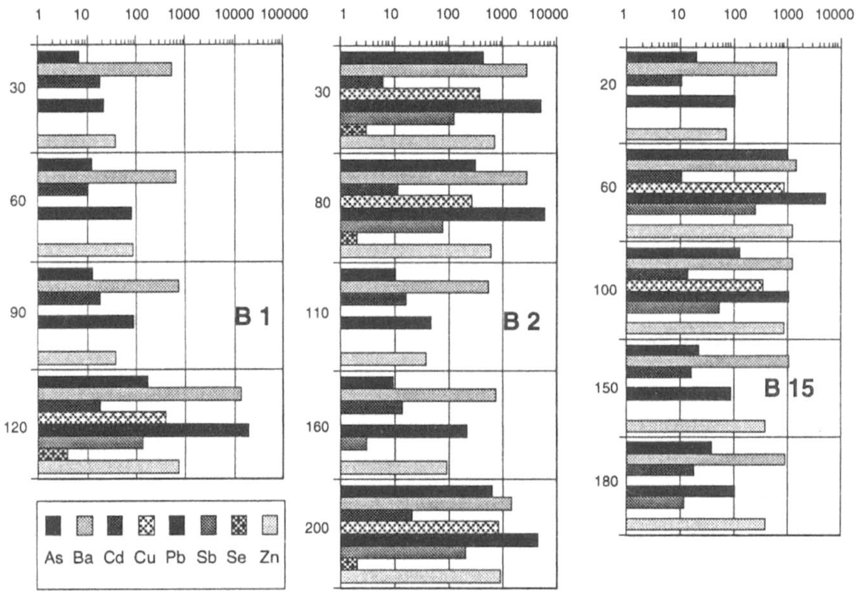

Fig. 7.2. Three soil profiles in horizontal alignment over the contaminated site (example from a total of 14 profiles; y-axis: depth given in cm, x-axis: log conc. given in μg/g).

7.5 Conclusions

The presented results, including the generally high heavy metal contamination of the area (soils and stream sediments) both up- and downstream from the lead processing factory, suggest that a disposal of some 1000 m^3 of contaminated soil would neither alter the general situation nor help the future perspective of the area. Instead, this disposal would take up valuable storage space on a disposal site and transfer the problem to another place. It was, thus, proposed to either excavate the contaminated soil (if room is needed for extensive foundations) and relocate it within the factory property on a prepared, slightly elevated site with appropriate drainage control and cover it with an asphalt seal to minimize water throughput. A second option would leave the contaminated soil in place and cover the site with concrete. To further extend the question, whether or not emissions from the lead processing factory have contributed to the pollution, more analytical work, e.g., XR-diffractometry (primary and secondary minerals in the soil) and mass spectrometry (Pb-isotopes) would have to be implemented.

Acknowledgements. The author thanks Dr. B. Knipping (†) and K. Herrmann for their practical support and advice with the measurements and optimization of the XRF-spectrometer system and Prof. emeritus A.G. Herrmann for the use of his laboratory facilities (all from the Department for Salt Deposits and Underground-Repositories, Institute for Mineralogy and Mineral Resources, Technical University of Clausthal, Germany). The generous agreement by the management of the lead processing factory to publish these data is highly appreciated.

References

Bowen HJM (1979) Environmental Chemistry of the Elements. Academic Press, London,

Govindaraju K (1989) Compilation of working values and sample description for 272 Geostandards. Geostandards Newsletter. International Working Group "Analytical Standards of Minerals, Ores, and Rocks", Vandoevre-lès-Nancy XIII, 113 S.

Hiller F, Hartinger L, Kiehne H-A, Niklas H, Schiele R, Steil H-U (1990) Die Batterie und die Umwelt. (Kontakt & Studium) Expert Verlag, Esslingen

Kinner UH, Kötter L, Niklauß M, Dodt J (1986) Branchentypische Inventarisierung von Bodenkontaminationen - ein erster Schritt zur Gefährdungsabschätzung für ehemalige Betriebsgelände. Umweltbundesamt, Forschungsbericht 107 03 001, UBA-FB 86-016, Erich Schmidt Verlag, Berlin

Klärschlammverordnung AbklärV (1992) Bundesgesetzblatt, Teil I: 912–934, Bonn 28.04.92

Matschullat J, Niehoff N, Pörtge K-H (1991) Zur Element-Dispersion an Flußsedimenten der Oker (Niedersachsen); röntgenfluoreszenz-spektrometrische Untersuchungen. Z Dtsch Geol Ges 142:339-349

Nowak H, Preul F (1971) Untersuchungen über Blei- und Zinkgehalte in Gewässern des Westharzes. (Geol. Jahrb. Beih.) Hannover

Umweltbundesamt (ed)(1994) Daten zur Umwelt 1992/93. Erich Schmidt Verlag, Berlin, 688 pp

Wedepohl K-H (1995) The composition of the continental crust. Geochim Cosmochim Acta 59,7:1217–1232

8 The Acid Lakes of Lignite Mining District of the former German Democratic Republic

M. Schultze and W. Geller

8.1 Introduction

Large areas of central and eastern Europe have reserves of hard and brown coal. The map in Fig. 8.1 (following page) presents the according mining areas of Germany, Poland and the Czech Republic. The reserves of lignite in Germany with 56 x 10^9 tons of potential output correspond to 10% of the world reserves and >50% of the reserves of Europe (Möhlenbruch and Schölmerich 1992). There are two major lignite districts in the area of the former German Democratic Republic (G.D.R.): the central German district around the city of Leipzig and the Lusatian mining area in the surroundings of Cottbus, which both have been exploited since the 17th century. This exploitation culminated after World War II, when 90% of power generation in 1988 – the year of maximum output – was based on brown coal: globally the highest proportion of this kind of primary energy source (Couch 1992; Eißmann 1994). As a consequence, in the area of the former G.D.R., the per-capita emission of CO_2 from the total of primary energy sources was ranked second after the United States at 18.5 tons per year and per inhabitant (based on data from 1990); the respective per-capita emissions were 10.8 tons CO_2 for the area of the former West Germany (BMU 1994). The lignite output of 310 million tons was accompanied by 1353 million tons of overburden and by 1720 million m^3 of groundwater which had to be removed in 1988 (Bilkenroth and Koziol 1990). The output of lignite declined after the reunification of Germany (see Table 8.1; Möbs and Maul 1994) leading to a brown coal proportion of 69% of all energy sources in 1990 and to a brown coal reduction by about two thirds from 1988 to 1993 for the area of the former G.D.R. (von Bismarck 1993).

The areas of open-cast mines were reclaimed after closure of the mining activities, and by 1992 up to 382 km^2 were recovered in the Lusatian district. Of

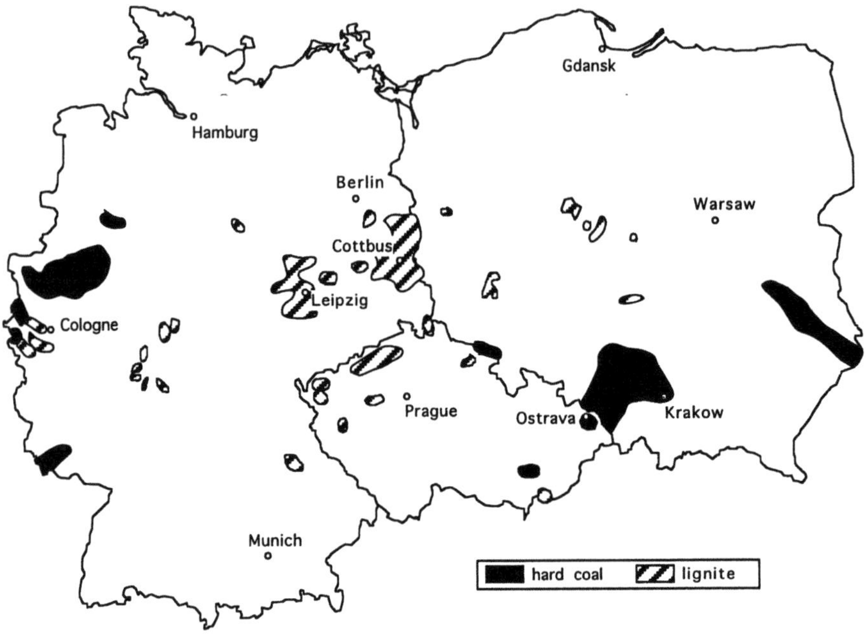

Fig. 8.1. Coal reserves of central Europe – in Germany (the central German district around Leipzig, and the Lusatian district at Cottbus), in the Czech Republic and in Poland. *Black areas* represent hard coal; *hatched areas* represent lignite (redrawn after Pätz et al. 1989)

this total of 382 km², 226 km² became forests, 87 km² arable farmland, and 36 km² lake areas (Henning 1994). The areas of lowered groundwater level have a considerable extent of 1100 km² in central Germany and 2100 km² in the Lusatian

Table 8.1. Output of brown coal and the number of active surface mines in the two lignite mining districts of the former German Democratic Republic (East Germany) between 1963 and 1993 (after Möbs and Maul 1994)

	Central-German district Lignite output in 10^6 t/y	No. of mines	Lusatian district Lignite output in 10^6 t/y	No. of mines
1963	145.5	–	108.7	17
1985	115.4	20	196.8	17
1988	109.8	20	200.5	17
1989	105.6	20	195.2	17
1990	80.9	19	168.0	16
1991	50.9	11	116.8	12
1992	36.3	9	93.1	9
1993	28.5	5	87.5	9
plan 2000	<20	3	70	5

district. Filling a groundwater body to its natural level would need 5.5 km^3 of water in the central German lignite district and 9 km^3 in the Lusatian district. In these two districts, water volumes of 2.5 km^3 and 4 km^3, respectively, would be necessary to fill in the open-cast mining holes to create future lakes. The water deficits accumulate to a total of 21 km^3 throughout the mining areas of the former G.D.R. This volume equals the biannual outflow of Lake Constance, the largest German lake. The available water in the area of the former G.D.R., however, is much less: 17.4 km^3 in years of normal precipitation, and 30–50% of this value in dry years (Aurada 1984).

The local problems primarily arise from the lowering of the groundwater level and from the oxidation and weathering of sulfuric minerals. Considerable portions of the overburden contain Fe-sulfides, pyrite and marcasite, which are oxidized as soon as the layers are removed and mixed, and thereby, aerated. The sulfur-acidic components, H_2SO_4 and Fe-hydroxides, are washed out by the rising groundwater reaching the respective levels and/or by leaching rainwater. Lastly, the acidic leachates are found downstream in the surface waters, and many of the mining lakes of the area are expected to become acidic after the filling of the open-cast lake basin. In a survey of the Lusatian mining district, 219 lakes of all sizes were investigated and about half of their number turned out to be acidic (Pietsch 1979).

8.2 Hydro- and Geochemical Processes Leading to Lake Acidification

The acidification of dumps and waters is a common problem in many mining areas. The according acidic mine drainage (AMD) has been investigated (e.g. Alpers and Blowes 1994; Fischer et al. 1987; U.S. Dept. of the Interior 1994). Oxidation of marcasite in dumps results in a first step in acid-reaction products which are then washed out. The most significant reactions of pyrite and marcasite oxidation are listed in Table 8.2 (after Stumm and Morgan 1981; Mattheß 1990).

Table 8.2. Most important reactions in acidification of lakes by oxidation of marcasite and pyrite (Stumm and Morgan 1981; Mattheß 1990)

(Eq. 1)	$2\,FeS_2 + 7\,O_2 + 2\,H_2O$	$\rightarrow$	$2\,Fe^{2+} + 4\,SO_4^{2-} + 4\,H^+$
(Eq. 2)	$FeS_2 + 14\,Fe^{3+} + 8\,H_2O$	$\rightarrow$	$15\,Fe^{2+} + 2\,SO_4^{2-} + 16\,H^+$
(Eq. 3)	$5\,FeS_2 + 14\,NO_3^- + 4\,H^+$	$\rightarrow$	$5\,Fe^{2+} + 10\,SO_4^{2-} + 7\,N_2$
(Eq. 4)	$FeS_2 + 14\,Fe^{3+} + 8\,H_2O$	$\rightarrow$	$15\,Fe^{2+} + 2\,SO_4^{2-} + 16\,H^+$
(Eq. 5)	$4\,Fe^{2+} + O_2 + 4\,H^+$	$\rightarrow$	$4\,Fe^{3+} + 2\,H_2O$
(Eq. 6)	$Fe^{3+} + 3\,H_2O$	$\rightarrow$	$Fe(OH)_3 + 3\,H^+$

Table 8.3. Acid-consuming reactions in solution, at soil–water interface, and used by microorganisms: carbonate buffering (Eqs. 7–9), ion exchange (Eq. 10), silicate weathering (Eqs. 11–12), gibbsite dissolution (Eq. 13), goethite dissolution (Eq. 14), denitrification (Eq. 15) and desulfurication (Eq. 16; Stumm and Morgan 1981, Mattheß 1990)

(Eq. 7)	$CO_3^{2-} + H^+$	→	HCO_3^-
(Eq. 8)	$HCO_3^- + H^+$	→	$H_2O + CO_2$
(Eq. 9)	$CaCO_3 + 2\ H^+$	→	$Ca^{2+} + H_2O + CO_2$
(Eq. 10)	$X\text{-}M + n\ H^+$	→	$X\text{-}H_n + M^{n+}$ [a]
(Eq. 11)	$KAlSi_3O_8 + 7\ H_2O + H^+$	→	$Al(OH)_3 + 3\ H_4SiO_4 + K^+$
(Eq. 12)	$2\ KAlSi_3O_8 + 9\ H_2O + 2\ H^+$	→	$Al_2Si_2O_5(OH)_4 + 4\ H_4SiO_4 + 2\ K^+$
(Eq. 13)	$Al(OH)_3 + 3\ H^+$	→	$Al^{3+} + 3\ H_2O$
(Eq. 14)	$FeOOH + H_2O + 3\ H^+$	→	$Fe^{3+} + 3\ H_2O$
(Eq. 15)	$5\ CH_2O + 4\ NO_3^- + 4\ H^+$	→	$5\ CO_2 + N_2 + 7\ H_2O$ [b]
(Eq. 16)	$2\ CH_2O + SO_4^{2-} + H^+$	→	$2\ CO_2 + HS^- + 2\ H_2O$ [b]

[a] X - material, capable to exchange ions, e.g. clay minerals, M - metal ions, n - value of charge;
[b] (CH_2O) - organic substances

In a follow-up process, reactions of bedrock weathering occur during seepage to the groundwater. There, portions of the solute may be reduced and intermediate minerals (siderite, goethite, jarosite, and gibbsite) may be precipitated (Ptacek and Blowes 1994; Bhatti et al. 1994; Cravotta 1994; Germain et al. 1994). The basic components are carbon dioxide, carbonic acid, and bicarbonate/carbonate (with reactions according to Eqs. 7–9), ion exchange of metals in acid solutions (see Eq. 10), weathering of alumo-silicates (see Eqs. 11 and 12), dissolution of Al- or Fe-minerals (see Eqs. 13 and14), and microbial reduction of NO_3^- and SO_4^{2-} with organic substances to CO_2 and to gaseous forms of N and S (Eqs. 15 and16; Mattheß 1990). These acid-consuming reactions are presented in Table 8.3.

The final quality of the lake water depends on the different reactions: the oxidation of marcasite and pyrite, and the secondary modification by weathering and reduction processes occurring during the seepage of the water through the ground.

8.3 State of the Mining Lakes in the former G.D.R.

The open-cast mining lakes in the area of the former G.D.R. show the full scale from neutral to very acidic lake waters. Especially in the Lusatian lignite district some, large lakes exist for about four decades, remaining very acidic without

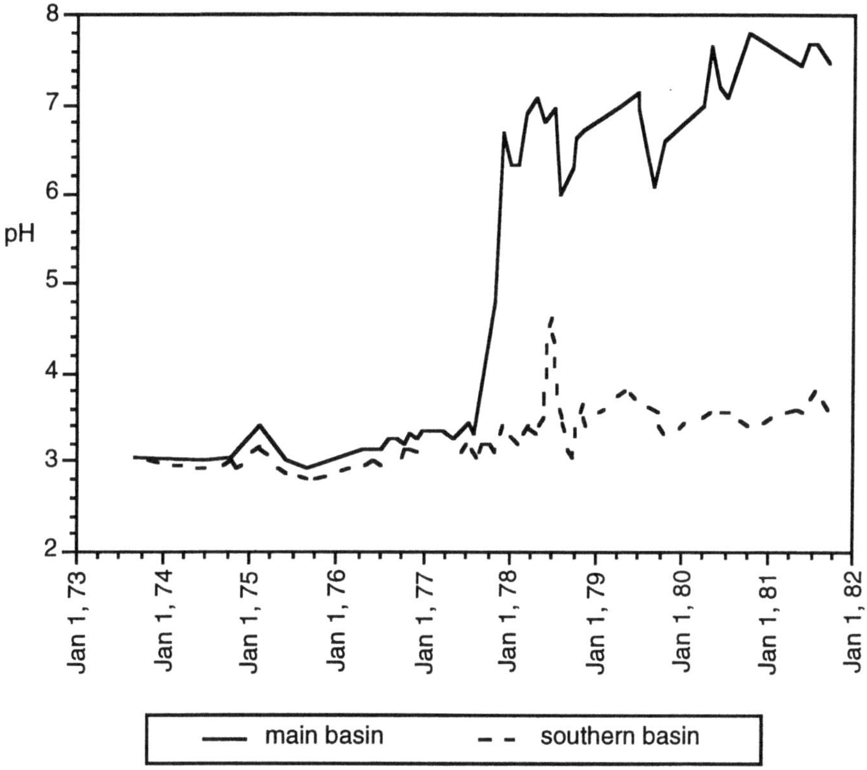

Fig. 8.2. Spatial distribution of pH-values in Senftenberger See during the years 1973–1981. Since 1976 there is a continuous throughflow of river water (Schwarze Elster) by which the main basin water is effectively exchanged, in contrast to the southern partial basin which is separated by an island and extended stands of macrophytes (redrawn after Pütz et al. 1991, as cited in Benndorf 1994)

major changes (Pietsch 1979). Some small lakes are well-protected against wind and, therefore, have a higher stratification stability, making possible the observed biologically mediated neutralisation in the hypolimnion. The well-mixed large and deep lakes, however, need an input of alkalinity from river water. An example is the mining lake, Senftenberger See, where the water of the main basin of the lake has been exchanged and is neutral. In a side basin the throughflow of riverine water is lacking and the pH-values remained strongly acidic (see Fig. 8.2).

In 1991–1993 a group of 21 open-cast mining lakes were investigated in a survey, and some were selected for more detailed research. The lakes are ranked in Table 8.4 according to their epilimnetic pH-values. These lakes can be grouped by the concentrations of Fe and Al. The most acidic waters with pH <2.7–3.4 show high Fe concentrations and lower Al concentrations. A second group of

Table 8.4. List of 21 open-cast mining lakes in the former G.D.R. depicting pH-values and concentrations of iron and aluminium as investigated in July 1993 (and [a] in September 1992)

Mining lake	pH	Fe (mmol l^{-1})	Al (mmol l^{-1})
Schlabendorf Nord F	2.69	3.04	0.76
Skado	2.7	2.69	0.55
Restloch 78	2.85	0.555	0.111
Sedlitz	2.9	0.824	0.471
Restloch 117	2.94	0.304	0.079
Halbendorfer See	3.0	0.358	0.073
Koschen	3.08	0.251	0.048
Waldsee	3.26	0.104	0.0085
Roter See	3.3	0.172	0.148
Cospuden	3.4	0.491	0.742
FKA Laubusch	3.75	0.030	0.102
Auenhain [a]	3.9	0.039	0.010
Rösa	4.0	0.012	0.052
Senftenberger See	6.71	0.0075	0.0007
Blauer See	6.8	0.0082	0.011
Seese West 4	7.22	0.0145	0.0015
Markkleeberg[a]	7.5	0.0215	0.0078
Grüner See	7.6	0.0034	0.0037
Schlabendorf Nord B	8.28	0.0016	0.0015
Hufeisensee	8.4	0.0018	0.0011
Hörlitz	9.05	0.0197	0.0337

lakes has intermediate pH-values with higher concentrations of Al than Fe, and the third group has pH-values near neutrality, with two to three orders of magnitude lower concentrations of both Fe and Al.

In Fig. 8.3, the measured pH-values of the investigated lakes are grouped by frequency. Two ranges of pH values appear in the recent investigation, one at pH 6 to 8 and a second at pH 2.5 to 4. Also seen from the literature (Pietsch 1979; Harvey 1980), a third pH range is revealed and a combined picture of all measured pH-values shows a three-modal frequency distribution. The ranges of pH-values and the concentrations of dissolved Al and Fe reflect the buffering potential of mainly three systems: the bicarbonate/carbonate system, the $Al(OH)_x$ system, and the $Fe(OH)_x$ system.

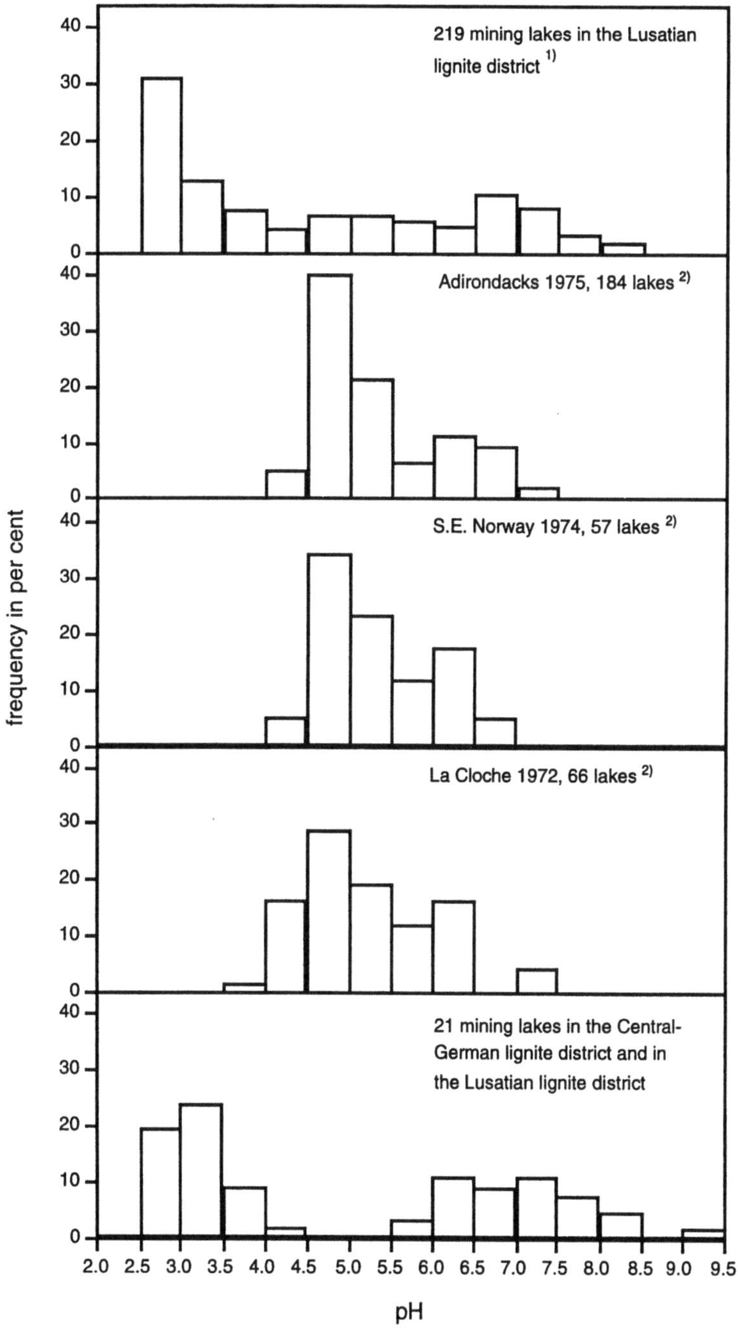

Fig. 8.3. Frequency distribution of pH in 21 surface-mining lakes in the former G.D.R. (Table 8.4), with two values from each lake representing the epi- and hypolimnion compared with other investigations (redrawn after [1] Pietsch 1979 and [2] Harvey 1980)

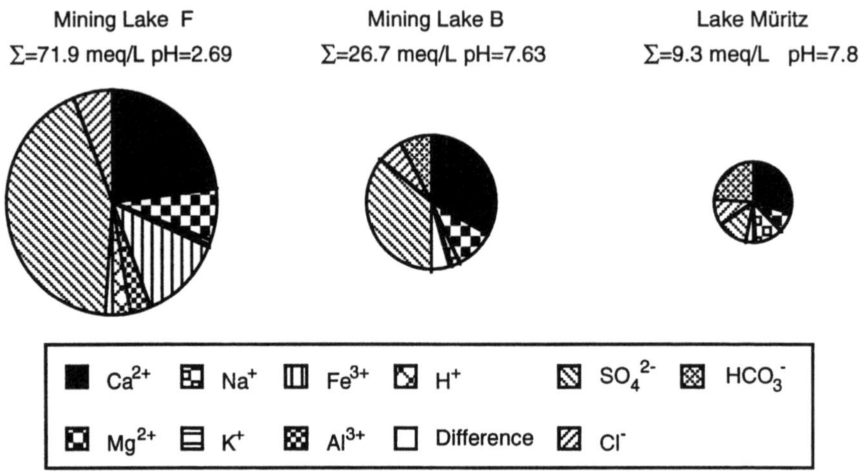

Fig. 8.4. Ionic composition in two closely related mining lakes, Schlabendorf Nord F and B, and in the largest glacial lake of the former G.D.R., Lake Müritz

The frequency distributions of pH as shown in Fig. 8.3 and the ionic composition of three representative lakes (see Fig. 8.4) demonstrate a considerable input of sulfuric acid also into neutral waters, where the buffering process is resulting lastly in increased levels of salt content, after reactions with carbonate and with silicate minerals of soil and weathering bedrock. By exhaustion of these initially reacting buffers, the pH-values decrease sharply, and with subsequent acidic conditions, Al and Fe become follow-up buffers. The Al-hydroxide system is already known to exist in soft-water lakes after rain acidification and was described in an extensive body of literature. The $Fe(OH)_x$ buffering system was known to be present in acidified soils (Ulrich 1981), but was not explicitly described to exist in lakes with pH 2–4.

The further mentioned buffering systems (see Table 8.3) as based on silicate and on ion-exchange appear to be of minor importance. Their capacity is presumed to be already exhausted in the catchment of the respective acidic lakes and, in contrast to groundwaters, the surface area of suspended particles and the amount of organic substances is negligible in the water body of the considered mining lakes.

The observed relationship between the pH-value and the respective alkalinity/acidity – as measured by titration to pH 4.3 – is presented in Fig. 8.5, as well as the resulting equivalent concentrations of dissolved Al and Fe. The *left panel* of Fig. 8.5 gives a survey of pH and alkalinity of the total number of 21 lakes. The graph resembles a titration curve from high to low pH-values – or vice versa – with two ranges of pH stabilized by buffering systems between pH 6 to 8 and between 2 to 4, and an intermediate range of transition between pH 4 to 6, where the pH is rapidly shifting with any addition of an acid or base. The concen-

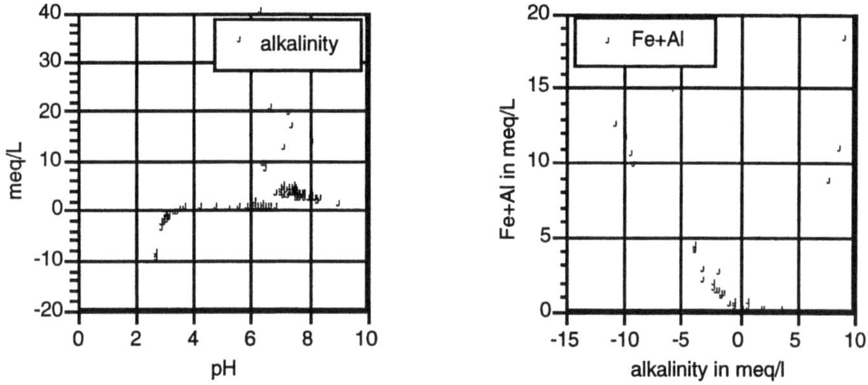

Fig. 8.5. *Left panel:* Values of alkalinity or acidity versus the pH-values of 21 mining lakes (see Table 8.4) as measured by titration to pH 4.3 with 0.1 *N* HCl, or with 0.1 *N* NaOH. *Right panel:* Fe + Al equivalent concentrations versus alkalinity or acidity, respectively, in 21 mining lakes (Table 8.4)

trations of Fe and Al shown in the *right panel* reflect the solubility of the respective minerals increasing with acidity by a linear slope of -1, and respectively demonstrate the dependence of the observed base capacity on the amount of dissolved species of Fe and Al as carriers of acidity. The same interpretation may be derived from Fig. 8.4, where the partial amount of sulfate, which is equivalent to the sum of Fe^{3+}, Al^{3+} and H^+, makes up the acidity according to the definition of Stumm et al. (1983):

$$\text{Acidity} = -(\text{Alkalinity}) = (SO_4^{2-} + Cl^- + NO_3^-) - (Ca^{2+} + Mg^{2+} + Na^+ + K^+)\,. \quad (17)$$

The high Fe-concentrations in the neutral range (see *right panel* of Fig. 8.5) refer to anaerobic conditions in the hypolimnion of some lakes where ferrous hydroxide is present.

The panels of Fig. 8.6a–e give an overview of the assembled values of 21 dissolved substances as measured through the survey investigation 1991–1993 in 21 mining lakes of the central Germany district and of the Lusatian lignite district. The samples were taken from different depths in the water column, reflecting the respective gradients of salt content and redox conditions that are found in the different lakes at different depths. The concentrations of the 21 dissolved substances are plotted versus the respective pH-value of lake and stratum.

The panel of Fig. 8.6a shows the major anions SO_4^{2-} and Cl^-, and the four major cations, Ca^{2+}, Mg^{2+}, Na^+, K^+. Only a few extremely acidic lakes have increased concentrations of sulfate, the bulk of all values is found in a range between 300 and 1000 mg l^{-1}. This reveals the pH to be less dependent on sulfate, but dependent on the kind and level of buffering substances to which the water seeping through the ground came into contact, thereby exchanging Al^{3+}-, Fe^{3+}-,

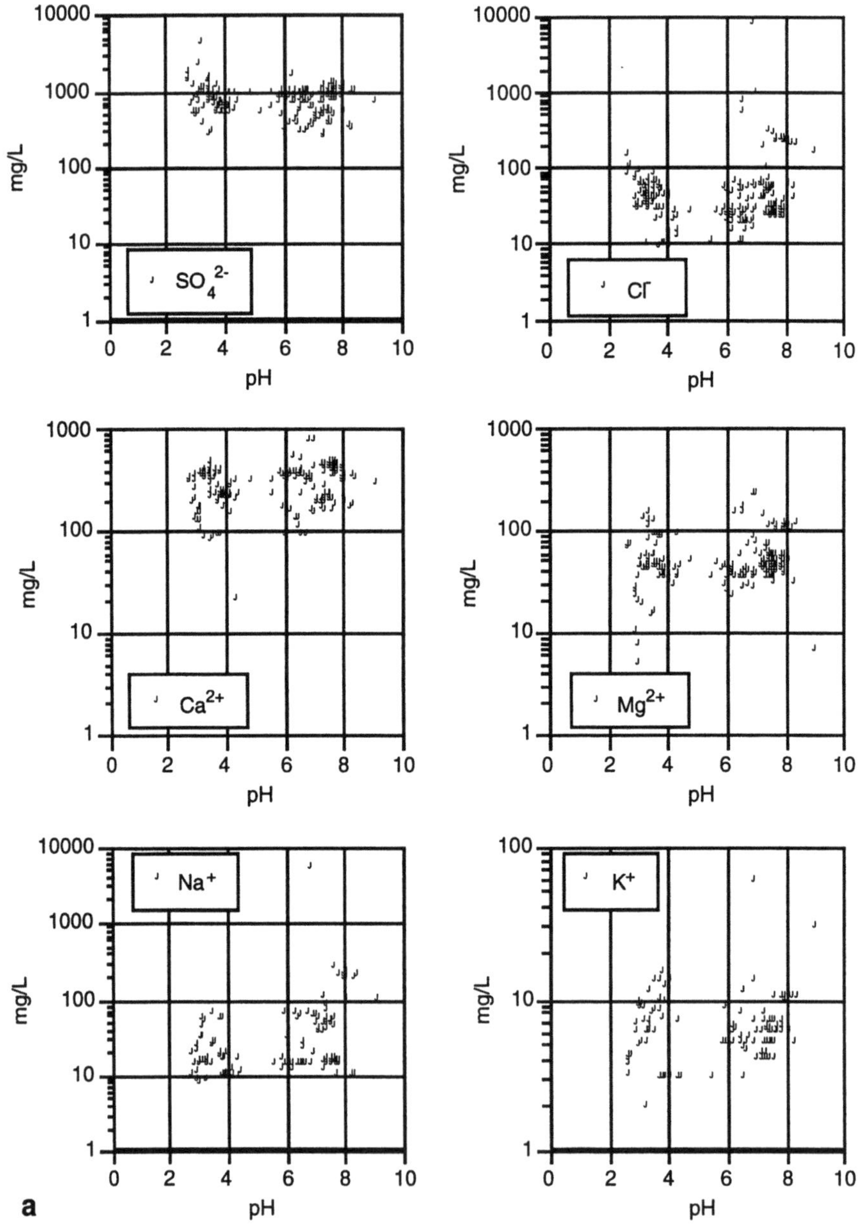

Fig. 8.6 a–e. 1991–1993 values of pH in 21 mining lakes of the former G.D.R. (see Table 8.4) versus the respective concentrations **a–b** of dissolved sulfate, chloride, 12 metals (including all speciation forms), **c** of Nitrogen as nitrate, ammonium and organically bound N, **d** of soluble reactive (SRP) and total Phosphorus (TP), and **e** of dissolved organic (DOC) and inorganic Carbon (IC). * represents values below detection limit

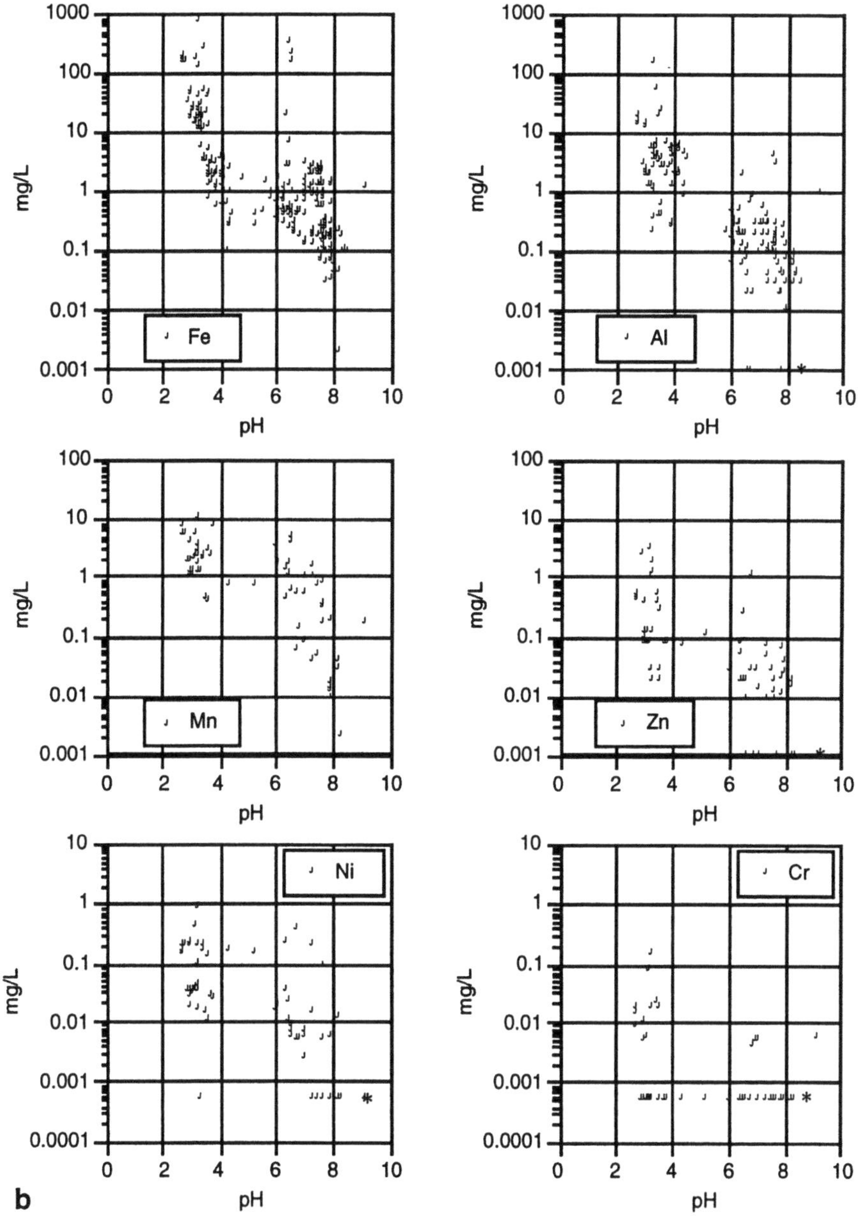

b

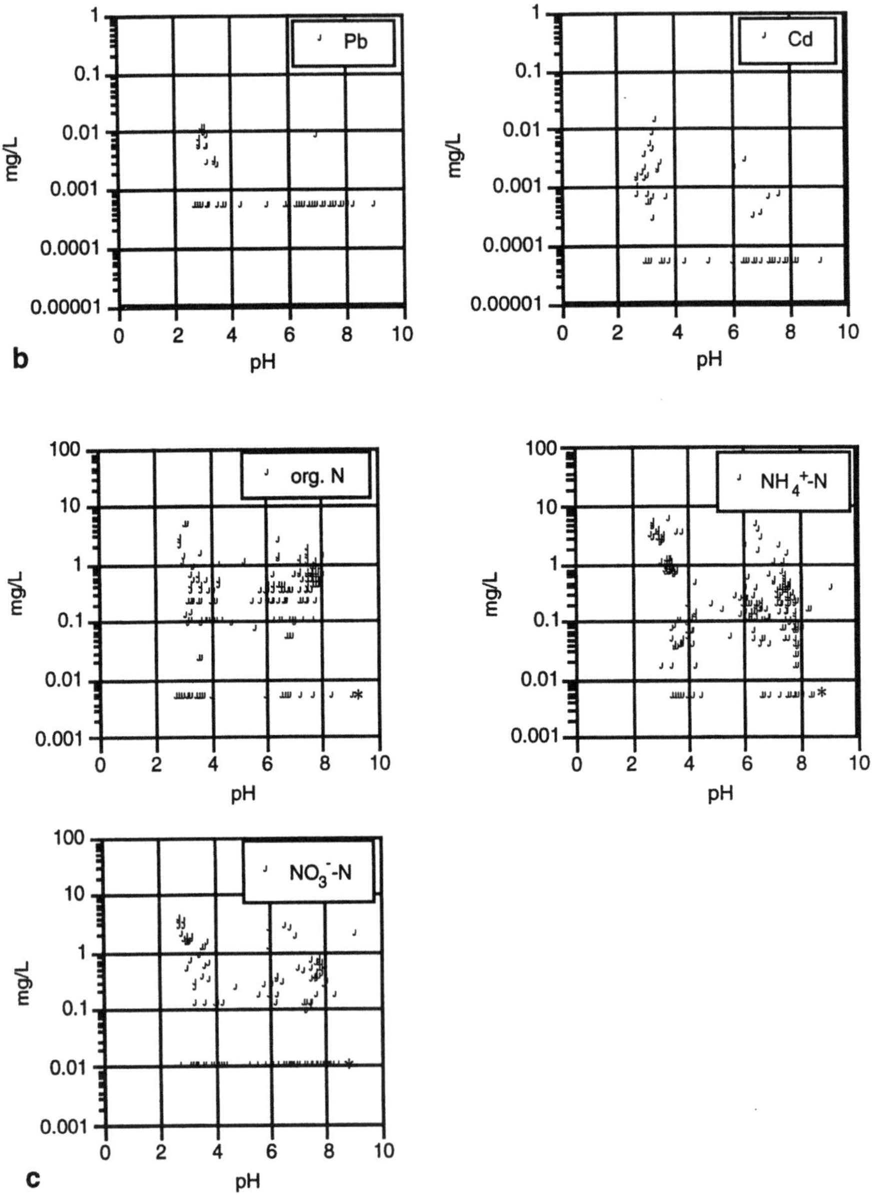

Fig. 8.6 b–c. continued

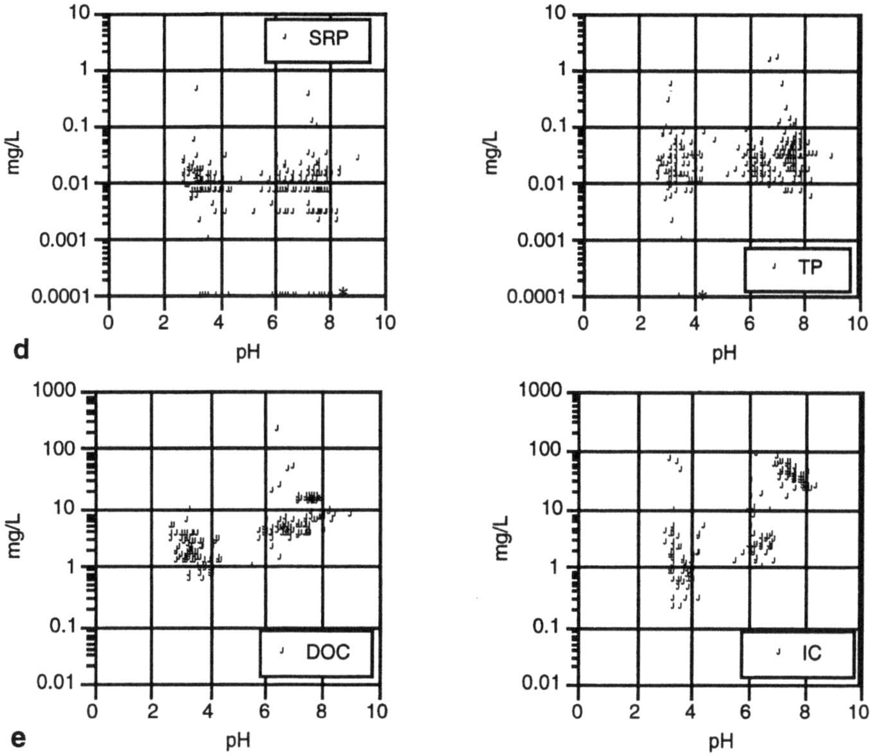

Fig. 8.6 d–e. continued

and H^+-ions with alkali cations, but leaving the mobile sulfate unchanged. Unusually high chloride concentrations originate mainly from Permian salt layers locally accompanying the coal. The same is true for high contents of Na^+ and K^+ in some cases.

The concentrations of Ca and Mg ions show no clear dependence on the pH, indicating buffering through soil passage after marcasite oxidation and subsequent leaching of the alkaline cations (see Table 8.3). The content of potassium in lake waters is near the groundwater concentration, indicating that these cations are effectively retained also under acidic conditions in soil and sediments (see also Table 8.5). In contrast, Fe, Al and six other heavy metals (see Fig. 8.6 b) show rather clear dependencies on the pH, mainly because of their strongly pH-dependent solubility, which is also demonstrated by their respective pK-values for hydrolysis (Table 8.6). The specific pK-values, however, are differing among the metals. Furthermore, these elements tend to be differently adsorbed by sediments or to be coprecipitated with neutralisation processes. The effect of changing redox conditions on Fe and Mn is also specifically different with high solubility of their

Table 8.5. Threshold-values of pH for the release of some heavy metals bound in sediment layers of the east Bavarian lake, Großer Arbersee (after Steinberg 1987)

Heavy Metal	pH
NH_4^+	5.5–6.0
Mg	6.0
Al	4.0
K	2.75
Ca	5.4–6
Mn	5.5
Fe	5.0
Cu	4.0–4.5
Cd	4.0
Pb	3.25

reduced forms under anaerobic conditions, explaining some selected high values of these elements in the range between pH 6 and 8.

Respective differences of solubility were also observed for the reverse process. Steinberg (1987, Table 8.5) gives some empirically observed threshold values of pH beyond of which ammonium and bound metals were released from the sediments of an acidified soft-water lake. The specific capability of metals to be exchanged by H^+-ions and, thereby, released from the sediments into the water is explained by the different radius and charge of the metal ions.

At any given pH-value, the ions show a range of concentrations in the lakes over two orders of magnitude. This kind of variability might be caused by locally divergent amounts of the respective elements in bedrock and soils through which

Table 8.6. pK-values of equilibrium constants for the first step of hydrolysis of some metals (after Stumm and Morgan 1981)

Reaction			pK
$Fe^{3+} + H_2O$	$\rightarrow$	$FeOH^{2+} + H^+$	2.19
$Al^{3+} + H_2O$	$\rightarrow$	$AlOH^{2+} + H^+$	4.97
$Zn^{2+} + H_2O$	$\rightarrow$	$ZnOH^+ + H^+$	8.96
$Cd^{2+} + H_2O$	$\rightarrow$	$CdOH^+ + H^+$	10.1
$Mn^{2+} + H_2O$	$\rightarrow$	$MnOH^+ + H^+$	10.6
$Mg^{2+} + H_2O$	$\rightarrow$	$MgOH^+ + H^+$	11.4
$Ca^{2+} + H_2O$	$\rightarrow$	$CaOH^+ + H^+$	12.85

the water was seeping. The plant nutrients N and P and their respective species show no pH-effect on concentrations (Fig. 8.6 c,d) according to the chemical nature of these elements. The organic and inorganic dissolved carbon demonstrates decreasing values with increasing acidic conditions. This reflects the decay of carbonate species – measured as inorganic carbon – by the stronger acids. High values are sometimes observed in an acidic hypolimnion, but here the CO_2 is presumed to originate from respiration processes in the sediment layers below.

There are several processes which might explain the decay or removal of organic substances from acidic water. High-molecular compounds like humic substances can be precipitated and , thereby, removed from the water. Productivity of algae is low, and low-molecular substances are released only in minor amounts, if any. Accompanying ferric hydroxide can be photoreduced by daylight and has been revealed to catalyze the oxidation and decay of DOC in acid-bog lakes (Colienne 1983).

8.4 Conclusions

The results of this and of other studies on mining lakes located in the area of the former G.D.R. show high concentrations of heavy metals if the water is acidic. In the case of neutralisation these toxicants are expected to be eliminated by precipitation. Bound in the sediments, they are no longer actually dangerous, but must be regarded as a potential source of intoxication.

The observed acidification of major proportion of several hundred lakes must be judged as highly artificial and far beyond a natural state. The consequence is that nearly all kinds of utilization of such surface waters are limited or completely excluded, e.g. fisheries. The aim of our research activities should, therefore, focus on the applied aspect on how to restore these lakes to a more natural state of ionic composition.

8.5 Measures and Approaches for Restoration

In the past, many acidic soft-water lakes have been restored by liming. The approach can be largely considered as a story of success (Brocksen and Wisniewski 1988; Olem 1991). The largest lake which was treated with a liming campaign was Lago d'Orta in northern Italy, where the acidic 1.3 km^3-volume of this seventh largest Italian lake was neutralized with more than 10 000 tons of $CaCO_3$ after about 60 years of industrial acidification (Calderoni et al. 1992). Liming, therefore, appears as a suitable measure for the treatment of soft waters with low ionic content which, however, are quite different from mining lakes. A first rough estimate for an artificial neutralisation of the acidic mining lakes of the Lusatian district reached twentyfold of the liming volume of Lake Orta: the necessity of 235 000 tons of limestone in a potential liming campaign was required and had to be distributed over the areas of a large number of lakes.

The easiest way to neutralize an acidic lake is presented by the example of Senftenberger See (Fig. 8.2; Benndorf 1994). Any throughflow of a river would effectively remove and wash out the acidic water. This washing out, however, would further induce a rather polluted lacustrine water causing other kinds of problems. Furthermore, considering the background of the known local water deficits, this approach is limited to few exceptional cases. A comparable way is to use groundwater pumped from active mines to the empty basins to be filled in closed-down mining areas.

The biological "production" of alkalinity appears as the best-working and cheapest method to reduce and to remove sulfuric acid, followed by a subsequent elimination of toxic metals from the water. The basic precondition is to reach an anaerobic environment to make possible the microbiological process of sulfate reduction. Given the presence of equivalent amounts of Fe, originating H_2S would be bound to the sediments by FeS precipitation. The basic philosophy of this approach, however, appears to be adverse to generally accepted guidelines of lake management to avoid eutrophication. We would initiate a process of nutrient increase by input of organic substances to stimulate anaerobic conditions in deepwater to reach a follow-up process of acid elimination. Perhaps this is a realistic approach, but many people might find it difficult to become accustomed to this concept.

References

Alpers CN, Blowes DW (eds) (1994) Environmental geochemistry of sulfide oxidation. American Chemical Society, Washington DC

Aurada KD (1984) Tendenzen der Gewinnung des verfügbaren Wasserdargebots in der mehrfach genutzten Landschaft. Petermanns geographische Mitteilungen, pp 115–125

Benndorf J (1994) Sanierungsmaßnahmen in Binnengewässern: Auswirkungen auf die trophische Struktur. Limnologica 24:121–135

Bhatti TM, Bigham JM, Vuorinen A, Tuorinen OH (1994) Alteration of mica and feldspar associated with the microbiological oxidation of pyrrhotite and pyrite. In: Alpers CN, Blowes DW (eds) Environmental geochemistry of sulfide oxidation. American Chemical Society, Washington DC, pp 90–105

Bilkenroth K-D, Koziol G (1990) Der Braunkohletagebau in der Deutschen Demokratischen Republik – Entwicklungsstand und Perspektiven. Braunkohle 3:5–10

Bismarck, F von (1993) Changes of German lignite mining: just the effects of reunification? In: OECD/IEA (ed) The clean and efficient use of coal and lignite: its role in energy, environment and life. Conf Proc, Hong Kong, pp 667–673

BMU (ed) (1994) Umweltpolitik. Erster Bericht der Regierung der Bundesrepublik Deutschland. Bundesministerium für Umwelt, Naturschutz und Reaktorsicherheit, Bonn

Brocksen R W, Wisniewski J (eds) (1988) Restoration of aquatic and terrestrial systems. Special issue of Water, Air, and Soil Pollution, vol 41, 1–4

Calderoni A, Mosello R, Ruggiu D (1992) Sixty years of limnology on Lago d'Orta: a case history of recovery from heavy pollution. Mem Ist Ital Idrobiol, 50:201–223
Colienne R H (1983) Photoreduction of iron in the epilimnion of acidic lakes. Limnol Oceanogr 28:83–100
Couch, G R (1992) Low rank coal in eastern Europe: opportunities and constraints. In: OECD/IEA (ed) Clean and efficient use of coal: the new era for low-rank coal. Conf Proc, Budapest, pp 243–260
Cravotta CA III (1994) Second iron–sulfate minerals as sources of sulfate and acidity: geochemical evolution of acidic ground water at a reclaimed surface coal mine in Pennsylvania. In: Alpers CN, Blowes DW (eds) Environmental geochemistry of sulfide oxidation. American Chemical Society, Washington DC, pp 345–364
Eißmann L (1994) Leitfaden der Geologie des Präquartärs im Saale-Elbe-Gebiet. Altenbg. nat. wiss. Forsch. 7:11–53
Fischer R, Reißig H, Peukert D, Hummel J (1987) Untersuchungen zur Beeinflussung der Markasitverwitterung. Neue Bergbautechnik 17:60–64
Germain MD, Tasse N, Begeron M (1994) Limit to self-neutralization in acid mine tailings: the case of East Sullivan, Quebec, Canada. In: Alpers CN, Blowes DW (eds) Environmental geochemistry of sulfide oxidation. American Chemical Society, Washington DC, pp 365–379
Harvey H H (1980) Widespread and diverse changes in the biota of North American lakes and rivers coincident with acidification. In: Drablos D and Tollan A (eds) Ecological impact of acid precipitation. Proc Int Conf, SNCF, Oslo
Henning D (1994) Bergbauliche Rekultivierung und Landschaftsgestaltung im Lausitzer Braunkohlerevier. Braunkohle 3:4–8
Mattheß G (1990) Die Beschaffenheit des Grundwassers, 2nd edn. Gebrüder Borntraeger, Berlin
Möbs H, Maul C (1994) Sanierung der Braunkohlegebiete in Mitteldeutschland und in der Lausitz. Wasserwirtschaft Wassertechnik 3:12–18
Möhlenbruch N, Schölmerich U (1992) Landschaften nach der Auskohlung. Spektrum Wiss 04/92:105–119
Olem H (1991) Liming acidic surface waters. Lewis Publ, Chelsea, Michigan
Pätz H, Rascher J, Seifert A (1989) Kohle – ein Kapitel aus dem Tagebuch der Erde. 2. Aufl., Teubner Verlag, Leipzig
Pietsch W (1979) Zur hydrochemischen Situation der Tagebauseen des Lausitzer Braunkohlen-Revieres. Arch Naturschutz Landschaftsforsch, Berlin, 19:97–115
Ptacek CJ, Blowes DW (1994) Influence of siderite on the pore-water chemistry of inactive mine-tailings impoundments. In: Alpers CN, Blowes DW (eds) Environmental geochemistry of sulfide oxidation. American Chemical Society, Washington DC, pp 172–189
Steinberg C (1987) Gewässerversauerung. In: Wagner R (ed) Wasser-Kalender 1987. Erich Schmidt Verlag, Berlin, pp 139–161
Steinberg C, Lenhart B (1986) Diskussionsbeiträge zur Geochemie der Gewässerversauerung. Deutsche gewässerkundliche Mitteilungen 30:1–9
Stumm W, Morgan JJ (1981) Aquatic chemistry, 2nd edn. J. Wiley and Sons, New York Chichester Brisbane Toronto Singapore
Stumm W, Morgan JJ, Schnoor JL (1983) Saurer Regen, eine Folge der Störung hydrogeochemischer Kreisläufe. Naturwissenschaften 70:216–223
Ulrich B (1981) Die Rolle der Wälder für die Wassergüte unter dem Einfluß des sauren Niederschlags. Agrarspektrum 1:212–231
U.S. Department of the Interior (1994) International land reclamation and mine drainage conference and third international conference on the abatement of acidic drainage. Vol 1–4: Mine Drainage. Proc Conf Pittsburgh 1994. Bureau of Mines Special Publ SP 06A/B/C/D–94

9 Metals in Lakes: Field and Modeling Results on Remedial Strategies with a Focus on Mercury and Radiocesium

L. Håkanson

9.1 Introduction and Aim

There are many chemical threats to the aquatic ecosystem today, e.g., eutrophication from nutrients, contamination from halogenated organics, radioisotopes and metals, and acidification. Pollutants reach lakes via different pathways: wet and dry atmospheric deposition on the catchment and the lake surface, direct emissions from, e.g., industries and urban areas and internal loading from contaminated sediments. The spread, retention and biouptake vary from substance to substance (Förstner and Müller 1974; Förstner and Wittmann 1981; Salomons and Förstner 1984; Vernet 1991), but the fundamental principles and processes regulating the distribution and biouptake are more or less the same for all toxic substances, and could be modeled in the same manner by means of generic, mass-balance models, but with different rates and model variables for different substances. For metals dissolved in the water or absorbed to very fine carrier particles, the distribution of these may be widespread and revealed as elevated concentrations in water, in suspended materials, in sediments and in biota over vast areas. However, elevated concentrations in abiotic compartments is one thing – biological uptake and increased ecological effects on the ecosystem level may be quite another (Håkanson 1984; Meili 1991a; Peters 1991). Thus, a fundamental question concerns the actual and potential ecological effects: How can one detect, describe and predict ecological effects at the ecosystem level?

It should be stressed that it is at least theoretically possible to model concentrations of substances in abiotic compartments like sediments and water in a rather straightforward manner. It is often much more difficult to apply causal models for biological variables and predict biouptake and concentrations in plankton and fish, and it is very difficult indeed to model, especially to have good, validated predictive models, for the really interesting issues, namely for the ecological effects on the ecosystem level. Elevated concentrations of contaminants that cause no visible or measurable ecological effects would generally be of less interest for practical water management and to remedial strategies in the situation faced today with multiple threats (acidification, contamination and eutrophication).

The aim of this chapter is first to present results from extensive field experiments in 95 Swedish lakes during 1986–1989 in order to reduce the concentrations of mercury and radiocesium in fish for human consumption (mainly pike and perch). The results from this project have been presented separately for mercury (Håkanson et al. 1990a; Lindqvist et al. 1991) and cesium (Andersson et al. 1991; Håkanson 1991; Håkanson and Andersson 1992). From the experiences gained from these field experiments, the aim here is also to discuss some *basic principles* concerning the distribution and biouptake of metals in general in lake ecosystems with a focus *on remedial strategies*. This will be done within a modeling framework and in this chapter a new model will be introduced. This model is based on both traditional mass-balance concepts used in dynamical/mathematical modeling and on concepts from empirical/statistical modeling (Håkanson 1991; Håkanson and Peters 1995).

9.2 Short Background on Mercury and Cesium in Swedish Lakes

The concentration of mercury in over 10 000 out of a total of some 83 000 Swedish lakes is deemed to exceed the "blacklisting limit" of 1 mg Hg/kg wet weight (related to 1-kg pike); about 40 000 lakes have fish (top predator, like pike) with Hg-concentration higher than the guideline value of 0.5. The levels are still rising in most lakes (Lindqvist et al. 1991). Mercury concentrations in soils are also continuing to rise, involving an increasing risk of toxic effects on the biological processes which influence metabolism in soil, among other things. Increasing mercury amounts in soil also means greater leaching of Hg, which in turn means that the load borne by lakes also increases. The long-term environmental goal set by the National Swedish Environmental Protection Agency (SNV) is that the level of Hg in fish must not exceed 0.5 mg/kg wet weight. This would mean that most people would be able to eat fish from lakes without needing to consider health risks.

The nuclear accident at Chernobyl resulted in large parts of Sweden being exposed to a major fallout of radionuclides in late April–early May, 1986 (Moberg 1991). In about 14 000 lakes (fall 1987) the fallout led to lake fish (1-kg perch) receiving concentrations far above the guidelines for commercial marketing of fish recommended by the National Food Administration (1500 Bq/wet weight; see Andersson et al. 1990). These consequences of the Chernobyl accident, together with the mass media debate, have implied a major economic setback for those parts of the tourist industry based on sport fishing, but have also caused problems for, e.g., fishery management and the local population who use lake fish as part of their basic diet (Bengtsson and Hansson 1990).

Many of the processes controlling the flow and biological uptake of Hg and ^{137}Cs in lakes are controlled by hydrological and morphometrical factors (Håkanson 1991), and thus are not influenced by remedies changing the water chemistry, e.g., liming and potash treatment. Remedies that could speed-up the recovery (meaning the reduction of the elements in fish) must aim at either blocking the biouptake, increasing bioproduction and causing a "biological dilution" of the elements in biota, and/or aim at adding specific elements, like potash (K) that can be taken up in fish in about the same way as ^{137}Cs and causing a "chemical" dilution (Black 1957; Fleishman 1963; Carlsson 1978), or elements like Se which would effectively bind Hg (Rudd and Turner 1983; Rudd et al. 1983; Björnberg et al. 1988; Paulsson and Lundbergh 1989).

There are some interesting principles for metals in general that could be addressed, using Hg and ^{137}Cs as type elements – e.g., the difference between a continuous contamination, like that for Hg, and the sudden peak in contamination for Cs, the fundamental differences between pelagic and benthic biouptake, the difference that Hg has an organic/lipophilic phase, which Cs has not, the importance of the partition coefficient and the differences in retention times between the two elements in biota.

The basic question in the "Liming–mercury–cesium" project was: What is the best way of reducing the concentration of these toxins in fish by means of liming and other ecologically acceptable measures? Metals can enter the fish along two main routes: (1) Direct uptake through gills and skin and (2) uptake via food. The food intake may be via the pelagic nutrient chain or the benthic chain; "you are what you eat" (Boudou and Ribeyre 1990; Meili 1991b). The amount of metals in fish (muscle, liver, kidney, etc.) depends not foremost on the total amount of the metal in the water mass but on how much of the metal that is bioavailable to the fish and its feed organisms. Bioavailability can, in turn, be regulated by a number of chemical and physical factors, which will be discussed in this chapter.

Methods of sampling, analyses and data handling, on descriptions of the lakes and their drainage areas are only briefly described in this chapter, since the aim here is not to provide a literature review.

9.3 Project Setup

Neither mercury nor cesium are known to pose any threat to the lake ecosystems. The problems concern the threat to humans eating contaminated fish. So, the concentrations of these toxic substances in fish muscle consumed by man will be called an effect parameter. Data have been collected with regard to the *effect* (concentration of Hg and ^{137}Cs in fish), the *dose* (load and/or concentration of Hg and ^{137}Cs on the lake), the *sensitivity* (different lake properties that could influence the biouptake) and the *remedy*.

The lakes were selected to yield as wide geographical and limnological ranges as practically possible. The cesium data emanate from 41 lakes in Västernorrland and Gävleborg county, and most of the discussions on Hg will focus on 75 lakes from all of Sweden for which we have data from the period 1986 to 1989. Considerable resources were used in all phases for sampling strategy, sampling, sample preparation, storage, analysis and analytical controls to establish the representativity and reliability of the data (Håkanson et al. 1990b).

The lake dose of Hg has been determined in several ways by:

1. Monthly analyses of Hg in lake water samples using so-called "clean" techniques (Lindqvist et al. 1991).
2. Analysis of the Hg-concentration in sedimenting material in sediment traps placed in the lake (May–August).
3. Hg analysis of surface sediment data (0–1 cm; three samples per lake from different water depths).

The following lake variables were determined (monthly samples from central lake sites) for pH, Secchi depth, temperature, alkalinity, total-P, chlorophyll-a, conductivity, Fe, Ca, hardness (= Ca+Mg concentration) and color.

- Theoretical turnover time of the lake water and theoretical water discharge (as determined from the so-called specific runoff).
- Morphometrical data, e.g., volume, lake area and mean depth from bathymetric maps.
- Data on the catchment area, e.g., mire percent, soil types, rock types, land use, etc.
- Hg-determinations were made on pooled samples of muscle from 1+ perch (the term "1+ perch", referred to throughout this chapter, indicates perch fry that is $\geq$ 1 year old but $<$ 2 years old). The perch were caught in late July, August and September. The pike were caught during January–May. Weight/length standardized data on Hg-concentration for both pike (1-kg pike; from regressions according to Johnels et al. 1967) and perch (1+ perch) have been used.
- The deposition of ^{137}Cs has been calculated from the fallout map.
- Analyses of ^{137}Cs were made on unfiltered lake water samples from 15 lakes; 1–5 samplings/year during the ice-free period. In this way, a picture is also obtained of the variation in the lake load with time.

- The concentration of ^{137}Cs was determined in material from sediment traps placed in the lake during the production period in the deepest part of the lakes at two different levels (about 2 m beneath the surface and about 2 m above the bottom). This was done during 1986–87 in 15 lakes and during 1988–89 in 41 lakes.

- Analyses of ^{137}Cs in surface sediments (0–1 cm) were made on samples from 15 lakes on two occasions, autumn 1986 and autumn 1989.

Basic data on the lakes are given in Tables 9.1 and 9.2. These tables give statistical compilations of some relevant dose, effect and sensitivity parameters. Most lakes are smaller than 1 km^2; the largest lake is about 5 km^2, the smallest 0.03 km^2. The mean depth varies from about 1 m to about 10 m with a mean of about 4 m. The mean annual value for lake pH (pH 12; data from 12 months) before the remedies varied from 4.8 to 6.6; the mean pH value for all lakes was 6.0. The mean annual lake color (col12) also varied within wide limits (from 14 to 340 mg Pt/l). Most of the lakes have low alkalinity and a color value around 100, which is typical for Swedish forest lakes. The mean annual lake concentrations of total-P varies between 3.0 and 35 μg/l. The lakes demonstrate a fairly large spread with regard to Hg in water, sediment trap material and surface sediments. Before the remedies, the Hg-concentrations in fish throughout were very high (Hg in 1-kg pike is given by $Hg_{pi}=1.19$ mg/kg on average, see Table 9.2). The Hg-concentrations in 1+ perch (Hg_{pe}) varied considerably less than Hg_{pi}. In most of the lakes, the pike had higher Hg-concentrations than the Swedish blacklisting value of 1 (mg Hg/kg ww); in some lakes the pike had Hg-concentrations above 2.

There are large differences in analytical reliability for different parameters (Håkanson et al. 1990b). Mercury can generally be determined with comparatively great analytical accuracy in fish and sediments. The average relative standard deviation (V in %) is only about 2–3% for Hg and about 5% for ^{137}Cs (Håkanson 1991). The pH can also be determined with great reliability. The average V value for pH (which by definition is a logarithmic value) is about 2%, and this includes all phases from sampling, sample preparation and analysis. Parameters that can be determined with fairly good reliability are, e.g., conductivity, hardness (CaMg) and Ca. Color (Col), Fe, total-P and alkalinity give considerably higher V values.

Table 9.1. Statistical compilation (n, min., max., median, mean *MV*, standard deviations *SD* and relative standard deviations $V = SD \cdot 100/MV$), for the "Cs-lakes". (Data from Andersson et al. 1991)

	Cs_{soil} kBq/m²	Cs_{wa}86-mod Bq/l	Area km²	Dm m	T year	ADA km²
n	41	15	41	41	41	41
Min.	2.5	0.04	0.07	1.1	0.02	1.0
Max	70.0	12.9	2.70	10.1	2.90	80.0
Median	15.0	0.9	0.40	4.0	0.47	17.0
MV	25.2	2.2	0.68	4.1	0.72	18.5
SD	21.1	3.3	0.64	2.2	0.68	16.5
V	83.8	148	94.5	53.3	95.12	89.1

	pH	cond mS/m	color mg Pt/l	totP µg/l	K µeq/l
n	41	41	41	41	23
Min.	5.1	1.6	35	4.3	5.9
Max	6.6	8.2	201	26.4	36.7
Median	6.1	2.6	109	10.0	8.7
MV	6.0	3.0	101	10.5	11.1
SD	0.4	1.2	41	4.2	7.3
V	6.4	39.6	40	40.5	65.4

	Cs_{pi}87 Bq/kg ww	Cs_{pi}88	Cs_{pi}89	Cs_{pi}86 Bq/kg ww	Cs_{pi}87	Cs_{pi}88	Cs_{pi}89
n	41	41	41	41	41	41	41
Min.	195	304	241	346	103	72	47
Max	7310	13756	20455	104809	19580	7641	3753
Median	1355	2497	2463	3380	2120	682	494
MV	**2392**	**3977**	**4768**	**8230**	**4066**	**1599**	**883**
SD	2260	3846	5185	16538	4877	1979	960
V	94	97	109	201	120	124	109

For Table 9.1: Cs_{soil}= fallout of ^{137}Cs after Chernobyl, $Cs_{wa}86\text{-}mod$= model calculated data on ^{137}Cs in lake water (Håkanson 1991), *pH* = mean annual lake pH (before the remedial measure), *cond*= mean annual conductivity, *color*= mean annual lake color, *totP*= mean annual total-P, *K*= mean annual K-concentration, Cs_{pi} = ^{137}Cs concentration in pike and Cs_{pe} = ^{137}Cs concentration in perch fry (= 1+ perch)[a]

For Table 9.2 (following page): *pHt0*= mean annual lake pH before the remedial measure, *pH88*= mean annual lake pH in 1988, *cond*= mean annual conductivity, *color*= mean annual lake colour, *totP* = mean annual total-P, *THg*= mean annual total-Hg in lake water, *RHg*= mean annual reactive-Hg in lake water, *IHg*= mean annual nonreactive-Hg in lake water, *RIHg*= mean annual reactive-Hg plus nonreactive-Hg in lake water, *SHg*= mean Hg-concentration from surficial (0–1 cm) sediment samples, $Hg_{pi}t0$= Hg-concentration in pike from the t0-period (= before the remedies) and Hg_{pe} = Hg-concentration in perch fry (= 1+ perch)[a]

For both Tables 9.1 and 9.2: *Area*= lake area, *Dm* = mean depth, *T*= theoretical lake water retention time, *ADA* = area of drainage area, [a] perch fry that is ≥ 1 year old but < 2 years old

Table 9.2. Statistical compilation (n, min., max., median, mean *MV*, standard deviations *SD* and relative standard deviations $V = SD \cdot 100/MV$), for the "Hg-lakes". (Data from Håkanson et al. 1990a)

	Area km²	**Dm** m	**T** year	**ADA** km²		
n	75	75	75	75		
Min.	0.03	0.9	0.02	0.4		
Max	5.26	10.1	4.79	80.1		
Median	0.40	3.5	0.53	12.1		
MV	0.75	4.2	0.89	15.6		
SD	0.89	2.5	0.98	14.7		
V	118	59.7	111	94.0		
	pHt0	**pH88**	**pH89**	**cond** mS/m	**color** mg Pt/l	**totP** μg/l
n	75	75	75	75	75	51
Min.	4.8	5.3	5.7	1.6	14.0	3.0
Max	6.6	7.4	7.3	10.8	340.0	34.7
Median	6.0	6.6	6.6	3.1	95.0	10.0
MV	**6.0**	**6.6**	**6.7**	4.1	99.2	10.8
SD	0.4	0.3	0.3	2.4	54.8	5.7
V	6.6	5.2	4.3	57.9	55.2	53.0
	THg ng/l	**RHg** ng/l	**IHg** ng/l	**RIHg** ng/l	**SHg** ng/g dw	
n	25	25	25	25	25	
Min.	1.9	0.6	0.4	1.0	156.2	
Max	9.4	2.3	3.4	4.9	461.7	
Median	5.1	1.4	1.8	3.4	226.7	
MV	5.0	1.4	1.9	3.3	237.6	
SD	1.8	0.4	0.7	1.0	71.9	
V	35.5	31.3	38.2	30.5	30.3	
	Hg_{pi}t0	**Hg_{pi}88** mg/kg ww	**Hg_{pi}89**	**Hg_{pe}t0**	**Hg_{pe}88** mg kg/ww	**Hg_{pe}89**
n	74	69	73	75	72	70
Min.	0.27	0.40	0.24	0.04	0.03	0.02
Max	2.23	2.19	2.21	0.45	0.40	0.32
Median	1.25	1.08	1.21	0.17	0.14	0.11
MV	**1.19**	**1.15**	**1.15**	**0.18**	**0.15**	**0.12**
SD	0.43	0.44	0.46	0.09	0.07	0.06
V	35.94	38.30	39.57	48.83	48.50	48.43

9.4 Remedies and Costs of Remedies

Different remedial strategies were tested in this project:

1. To reduce the transport of Hg and ^{137}Cs from the catchment area into the lake, the measures wetland liming and drainage area liming were tested.
2. To change the foodweb so that the concentrations decrease in fish. This applies to intensive fishing (Göthberg 1979) and nutrient addition (foremost of phosphorus in low-productive lakes; see Håkanson 1980; Andersson et al. 1989). These measures are intended to stimulate the primary and secondary production. However, to make a detailed study of the flow of Hg and ^{137}Cs in nutrient chains has been outside the scope of this project .
3. To reduce the bioavailability of the lake dose.

Lake liming. The durability of a liming input depends mainly on the dose of lime and the lake water turnover time (from Håkanson et al. 1990a; see Table 9.3).

Three different types of lime, bedrock lime, sedimentary rock lime and a special "mixed" lime containing several trace elements in addition to selenium and phosphorus were tested (see Håkanson et al. 1990a for recipe).

Lake liming + potassium: i.e., lime with addition of potash containing potassium.

Lake liming + selenium: (Rudd and Turner 1983; Rudd et al. 1983; Björnberg et al. 1988). Two different ways of adding selenium were tested, one using mixed lime and the other according to the "Boliden method" (Björnberg et al. 1988; Paulsson and Lundbergh 1989). The Boliden method implies that selenium is sealed in rubber tubes placed in the lakes in nets. The selenium is successively released. If the selenium treatment is to have optimal effect, the pH of the lake water should be fairly high. All lakes treated with selenium were also limed.

Lake liming + fertilizing: i.e., liming and phosphorus treatment. In one lake, the addition of phosphorus was done with commercially purchased fertilizer (Osmocote®), in two lakes emissions from fish-cage farms (rainbow trout) were used. The remedy is intended to give both an increase in pH and a biological buffering (Jansson et al. 1981).

Lake liming + intensive fishing. This measure is intended to increase pH, reduce highly Hg-contaminated fish, and modify the fish population and predation pressure.

Table 9.3. Dose of lime and theoretical retention time of lake water

pH before liming	Dose of lime, g/m³ (0–0.2/0.5 mm, 50% CaO) (Theoretical lake water retention time, years)		
	<0.5 years	0.5–3 year(s)	>3 years
5.6–6.0	30	20	10
5.1–5.5	45	30	15
4.6–5.0	60	40	20
4.3–4.5	75	50	25
4.0–4.2	90	60	30

Wetland liming. Wetland liming implies a considerably improved duration compared to lake liming, better conditions for biological life, e.g., for chironomid larvae and ephemeroptera in streams and rivers, reduced transport of metals (e.g., Fe, Cu and Al) into the lake, and possibly also reduced the load of Hg and ^{137}Cs to the lake. The following lime doses in Table 9.4 were applied.

Table 9.4. Lime doses on wetlands

	Total dose tonnes/ha of catchment area	Lime dose on wet lands 2%t tonnes/ha	10% wet land tonnes/ha
S. Sweden (Götaland)	1.0	50	10
C. Sweden (Svealand)	0.5	25	5
N. Sweden (Norrland)	0.3	15	3

Wetland liming + potassium.

Wetland liming + selenium (selenium additive according to the Boliden method).

Drainage area liming. Four complete but small catchments were treated with dolomite.

Intensive fishing (without liming): The intensive fishing resulted in reductions of 2.6 to 17.6 kg fish per hectare. Mainly pike were removed, but also perch, roach and other species.

Lake liming was tested in most lakes. The project used 32 881 tonnes of lime, 320 tonnes of potash and 2701 tonnes of dolomite. The total remedial costs were 28.8 million SEK, the lime costs 22 million SEK. In all, the project had a turnover of 44 million SEK during the period 1985 to 1990. The spread in the costs was large for one and the same remedy depending on how the operation was organized and the size of the lake. Some measures, e.g., wetland liming, drainage area liming and intensive fishing may be extremely expensive. In nine cases, the cost was in excess of 1 million SEK per lake.

9.5 Results of the Remedial Measures

One reason why the results of the remedies are difficult to interpret is that they were carried out differently. Some were meant to have a long-term effect (e.g., wetland liming and drainage area liming), whereas others were meant to reduce the Hg-concentration in fish without altering the water chemistry (e.g., intensive fishing). Drainage-area liming with dolomite did not give particularly increased pH-values in the lakes even after 3 years. Certain water chemical variables, such as pH, can respond rapidly to a liming. But how long does it take before the Hg-concentrations decrease in pike? And how long does it take before the Hg-concentrations in perch reach a new stable level following a remedy? Questions like these are brought into focus in the following sections.

9.5.1 Cesium

Table 9.1 shows how the concentrations of ^{137}Cs in fish have changed from 1986 to 1989. The concentrations in perch fry have decreased markedly, whereas the values in pike have continuously *increased.* Is this a natural pattern? Or has this been influenced by the remedies?

A clear effect of a liming on water quality is a prerequisite for the remedy being able to achieve a reduction in the fish's concentration of radiocesium. The mean pH in the lakes increased from *6.0 to 6.7*, and also other variables directly linked to pH, e.g., hardness and alkalinity. The lakes treated with potash had a relatively greater increase in pH and alkalinity than the other lakes because many of the potash-treated lakes were also given an "over-dose" of lime. The potash treatment generally led to a marked increase in K-concentrations. Most of the treated lakes had K-concentrations > 20 μeq/l after the treatment as compared to < 10 μeq/l before the remedies. The results of the potash treatment, just like any other treatment, vary depending on the water retention time of the lakes; in lakes with short turnover times, the K-dose has been insufficient to give a long-term effect (see Andersson et al. 1991 for further details about the durability of the treatments in different lakes). The concentrations of total-P showed little change, so this remedy was not carried out effectively.

The concentrations of ^{137}Cs in both water and sedimenting material decreased significantly during the period. The decrease was particularly evident between 1986 and 1987 when the half-life for the activity in the sedimenting material was, on average, slightly more than 100 days. The continued decrease is much slower. Shallower lakes have a slower decrease in Cs-concentration than deeper lakes, which may be explained by a larger resuspension in shallow lakes. This implies that a large proportion of the highly contaminated material in 1986 is found in the water mass of shallow lakes where it is available for biouptake.

The decrease in perch can be described fairly well with an exponential function. This indicates that the biologically easily absorbable Cs-fraction in the sedimenting material decreases with time. Significant correlations with the decrease in concentrations in water and the sedimenting are obtained for the lake dynamic ratio, DR (r = -0.57; where DR = $\sqrt{area}$/Dm and Dm is mean depth) and the max. depth, D_{max} (r = 0.45); and among the water chemical variables for color (r = -0.35); other correlation coefficients, e.g., for pH (r = 0.23) and the water retention time, T (r = 0.17) are nonsignificant. Lakes with the slowest Cs-decrease in small perch during the period are relatively shallow and brown lakes. The magnitude of the continued decline will be foremost influenced by secondary loading from the catchment and the internal loading.

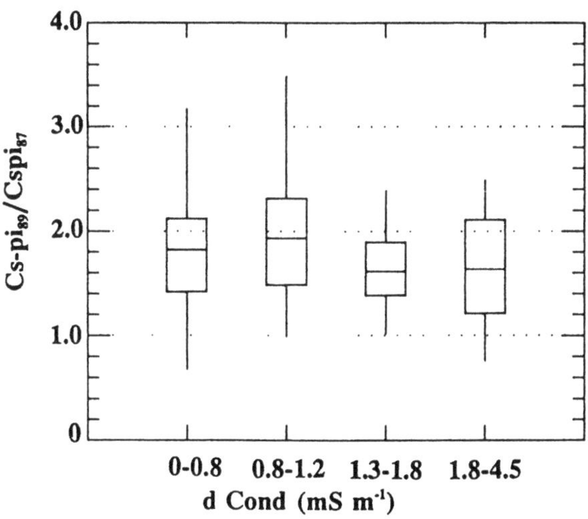

Fig. 9.1. Min., max. median and quartile values for the relative change in ^{137}Cs in pike from early 1987 (before remedies) to 1989 in relation to the change in lake mean conductivity (*dCond*; conductivity in mS m^{-1}) linked to the remedial measures. The lakes with the smallest difference in conductivity (i.e., dCond<0.8) may be regarded as "reference" lakes. (Andersson et al. 1991)

It is interesting to note that there is *no clear difference* in the decline in Cs-concentrations in fish between lakes where different remedies were tested. The magnitude of the decrease in Cs-concentration in for example perch, thus, must foremost be linked to other factors than the remedies. Despite ^{137}Cs in water, in sedimenting material and in perch fry, the concentrations in pike have increased during the period. Pike generally reaches almost as high levels as the highest levels measured in perch fry, but with a delay of 2–5 years.

From Fig. 9.1, one can note that the increased concentration in pike up to 1989 is slightly lower, about 5% (which is not significant) in lakes where the total ionic strength in water was increased by the remedies by more than about 1.5 mS/m. In a longer time perspective, however, a reduced Cs-uptake in pike of 5% per year may be of value (Andersson et al. 1991).

9.5.2 Mercury

Results for mercury are given in Fig. 9.2A for pike, where the values before treatments, $Hg_{pi}t0$, are compared to values 2 years after the treatments, $Hg_{pi}89$. One can note a significant relationship between the two parameters ($r^2 = 0.43$) and that all data are evenly spread around the line y=x. This means that on average *nothing* seems to have happened! The values have even increased in many lakes! The results were poor for ^{137}Cs, and the results in Fig. 9.2A suggest that the remedies had no significant effects on Hg in pike. So, have these extensive and expensive field experiments failed?

A first answer that question is given in Fig. 9.2B. This diagram shows, in fact, that much has happened, however, not for Hg in pike but for Hg in perch fry: There is a significant decrease in most lakes; in three lakes the values have decreased more than 75%, and in all 75 lakes the average decrease for all the measures is larger than 30%. It is evident that the retention of Hg in perch fry is much quicker than in large pike. An interesting general question is then: What is the ecological half-life of a metal in fish? And how can it be modeled?

The pike is generally caught during the spawning period in March/April. The Hg-content in pike depends on how the pike has lived and what it has eaten during a fairly long period before being caught. The Hg-concentration in pike is an integrated value for the entire environment of the pike and its prey. The Hg-content in pike has a certain half-life. If an Hg-contaminated pike were placed in a "Hg-free" environment, it would take about 3 years before the Hg-content had decreased to half its original level (Boudou and Ribeyre 1990; Meili 1991b). Thus, the Hg-content in pike depends on the Hg-dose supplied to the entire environment and not just to the clump of reeds where the pike was caught, depending as well on the biological, chemical and physical conditions in this environment over a long period, since these conditions influence the distribution of the Hg-dose on different carrier particles and the bioavailability of the Hg-dose. The same type of argument could be given for any metal or toxic substance. The biological half-life

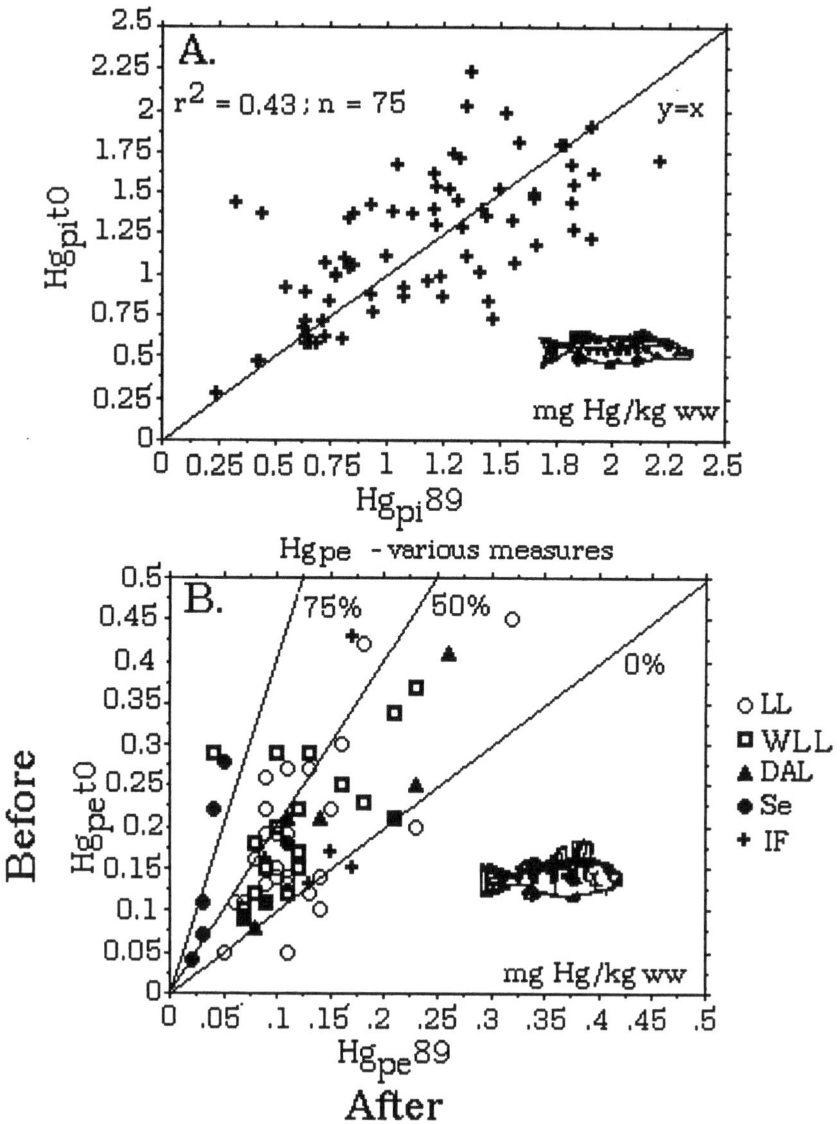

Fig. 9.2. A. The relationship between the Hg-content in 1-kg pike before remedial measure (carried out during the summer and autumn of 1987; given as $Hg_{pi}t0$) and the Hg-content in pike in 1989, i.e., 2 years after the remedies. The r^2-value is given and the number of lakes ($n = 75$). No changes mean that all data should appear on the *line* $y = x$
B. Corresponding plot for Hg-content in perch fry (1+ perch[a]; Hg_{pe}). Two lines for 50% and 75% reduction (before to after). *LL*=lake liming, *WLL*=wetland liming, *DAL*=drainage area liming, *Se*=liming plus selenium treatment, *IF*=intensive fishing. (After Håkanson et al. 1990a)

[a] perch fry that is ≥ 1 year old but < 2 years old

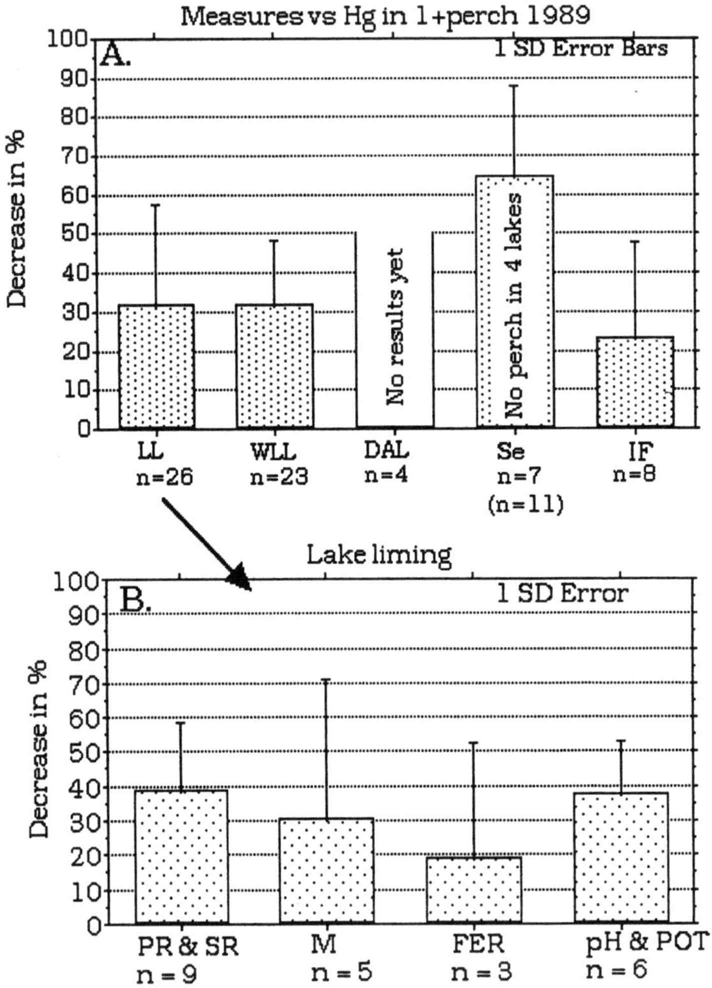

Fig. 9.3. A. Comparison of how the different remedies (*LL*=lake liming, *WLL*=wetland liming, *DAL*=drainage area liming, *Se*=selenium treatment plus liming and *IF*=intensive fishing) influenced the percentage change of Hg-concentration in *1+ perch*[a] before time of the introduced remedy (t0-data) and up to 1989
B. Comparison of how the different types of lime used in lake liming (*PR*=primary bedrock lime, *SR*=sedimentary rock lime, *M*=mixed lime, *FER*=fertilization, *pH & POT*=liming to higher pH using potash) influenced the percentage change Hg-concentration in 1+ perch[a] (Hg_{pe}) before the time of the introduced remedy (t0-data) and up to 1989. (After Håkanson et al. 1990a)
[a]perch fry that is ≥ 1 year old but < 2 years old

for Hg in pike is about 3–5 years and the biological half-life for Hg in small perch is about 1–2 years (Lindqvist et al. 1991).

Figure 9.3A gives a summary of the results for all tested remedies.

Intensive fishing . After 2 years, intensive fishing can reduce the Hg-concentration in 1+ perch by about 20–30%. Since pike eats perch, there is a causal link between Hg-concentrations in 1+ perch and Hg-concentrations in 1-kg pike. We may expect the Hg-concentrations in pike normally also would decrease by 15–40%, but this will take some more years. If the intensive fishing is to achieve this positive effect, it is necessary that this remedy is introduced efficiently (i.e., more than about 25% of the fish biomass has to be reduced). However, efficient intensive fishing is demanding both with regards to staff and economy, especially in lakes larger than about 1–10 km^2. It is probable that the effect of an intensive fishing normally has a duration of about 10 years, i.e., that it takes about 3–5 years before the Hg-concentrations in pike decrease to a minimum and then about 3–5 years for the concentrations to increase again. Intensive fishing is not a cost-effective remedy for Hg if current market salaries are to be paid to the people who conduct the fishing. However, it may be a suitable measure for fish protection societies, etc., who themselves wish to improve the Hg situation in a lake. The results of the intensive fishing could be improved in acidic lakes if this remedy is combined with liming.

Drainage area liming with dolomite was carried out in four catchments. This is an expensive remedy. It is done to obtain a long-term influence on the entire ecosystem (on the forest, soil chemistry, life in streams and lakes, and possibly also on the Hg-concentrations in fish). It was clear from the start of the project that it would not be possible within a period of 2–3 years to witness clear positive effects of this remedy with regards to Hg-concentrations in fish. After 2 years, one can note that the drainage-area limings have led to an increase in lake pH by about 8% on average. This remedy needs to be followed-up for several years.

Selenium treatment. It is important not to dose so that the Se-concentrations in water rise above about 1–2 μg Se/l. The Se-concentrations in fish should never be higher than about 2–3 mg Se/kg ww. If the selenium dosing can be managed in this manner, then the method should be interesting, particularly for large and severely Hg-contaminated lakes. There are, however, questions as to the efficiency of the Se treatment in reducing the Hg-concentration in fish if the Se is dosed in an ecologically acceptable manner (according to the above). In addition, there are questions concerning the effects of Se on different species of animals and plants that must be elucidated before the method can be recommended more generally. It is remarkable that the reproduction of perch decreased drastically in four of the eleven lakes treated with Se (the Boliden method). The data indicate that it is probably due to the Se treatment, but we have no information on the mechanisms that might be involved. The experience gained during the project with the Se method indicates that the dosing problem may be solved if the Se nets are placed near tributaries where the water turnover is fast, if the nets are not at central lake sites, and if the nets are only allowed to be in the lakes from spring until early autumn and not during periods with ice cover.

Lake liming. There is very good agreement between how the limings influence lake pH and the result as regards the Hg-concentration in 1+ perch. In a well-accomplished lake liming, it is possible to reduce the Hg-concentration in perch fry by about 30% in 2 years. The concentrations in pike may be assumed to decrease by 15–40% within a further 2–5 years. It is extremely difficult to elucidate how the lake limings influence the Hg uptake in fish in detail. It need not be the pH change that causally links to the Hg reduction in fish. It may be any of the factors influenced by the liming, e.g., the increase in Ca-concentration. It is not possible to see any difference between the different types of lime (bedrock lime, sedimentary rock lime and "mixed" lime) as regards the reduction in the Hg_{pe}-values (Fig. 9.3B). As a working hypothesis, it was assumed that "mixed" lime would be more effective than bedrock and sedimentary rock lime. However, this does not appear to apply. Since lake limings are conducted on a large scale in Sweden to minimize acidification damage (8000 out of 83 000 lakes have been limed; the total yearly cost exceeds 150 million SEK), it is important to note that lake limings can in fact also reduce Hg-concentrations in fish.

Wetland liming. About the same results are obtained for the Hg-reduction in perch fry with wetland liming as with lake liming. The most important aspect appears to be not whether use is made of bedrock lime, sedimentary rock lime, "mixed" lime or potash, but how efficiently the remedy is carried out. The Hg-concentration in perch fry can be reduced by about 30% in 2 years. The concentrations in pike may decrease by 15–40% within a further 2–5 years. Wetland liming does *not* appear to influence lake color or the Hg-transport from land to water to any particular extent (Håkanson et al. 1990a).

Fertlization. In this material, there is nothing to suggest that fertilization can reduce the Hg-concentration in perch to any particular extent. This need not exclude the possibility that lake bioproductivity can actually influence the Hg-concentration in fish in a positive manner ("biological dilution"), as has been indicated by other studies (Håkanson 1980), since these fertilizations were not carried out efficiently.

9.6 Dynamic Model - Setup

Dynamic models are based on a causal analysis of the processes governing the fluxes of matter and substances. If dynamic models are to be used in practice, e.g., to quantify fluxes of energy or contaminants in lakes, the rates that govern the transport between the various compartments in the lake ecosystem have to be known, simulated or estimated. Here, a new mixed dynamic model (Håkanson 1991; Håkanson and Peters 1995) will be introduced as a tool to discuss important principles to interpret the results of the remedial measures.

Figure 9.4 gives an outline of the model. The figure is meant to provide an overview of the fact that many factors and processes may affect the distribution and biological uptake of a metal in a lake. These factors are here grouped into four categories:

1. *Load* (= dose) factors; there are various processes causing high lake loads of metals, e.g., direct metal deposition to the lake surface, input of metal from the drainage area (from natural weathering and outflow of anthropogenic metals), and internal fluxes from sediments. There are many ways to define load variables for metals in water, suspended matter and bottom sediments, and many different chemical forms of metals (metallic, ionic, etc.)
2. *Biological* or *ecosystem* variables (like size and age of animals) related to the pelagic and benthic fluxes of metals in the foodweb, and
3. Environmental *sensitivity* factors linked to the morphology, hydrology and water quality of the ecosystem. These factors regulate the transport routes and biouptake rates of metals.
4. *Ecological effect parameters* should generally and ideally be measures of the reproduction or abundance of key functional organisms of the lake. *In this model, we simply use the metal concentration in top predator as the target effect parameter.*

All of the factors listed in Fig. 9.4 (and many more) could, potentially, influence the spread and biouptake of a metal. A central problem in predictive modeling is to *quantitatively rank the importance of such factors* and derive models that enable predictions of the y-variable, which in this example represents the metal content in top predator, from as few as easily obtainable and as relevant x-variables as possible.

In more traditional ecosystem modeling, it is often a goal to include as many relationships (fluxes and compartments) as possible and describe the complex relationships in a logical, causal manner. If, on the contrary, one would like to *predict about one or a few y-variables*, it is evident that it may be very impractical, costly and laborious to use very big ecosystem models. This mixed model has been constructed first of all to interpret the results presented for Hg and ^{137}Cs. It may also be used as a tool to simulate the spread and biouptake of metals in lake ecosystems more generally. The idea is *not* to include every possible process, but the crucial processes. Mixed models in this context are models that are based on both *dynamic* characteristics *and empirical* relationships (Håkanson 1991). There are several ways to link these two fundamentally different model approaches, and here we will give some examples.

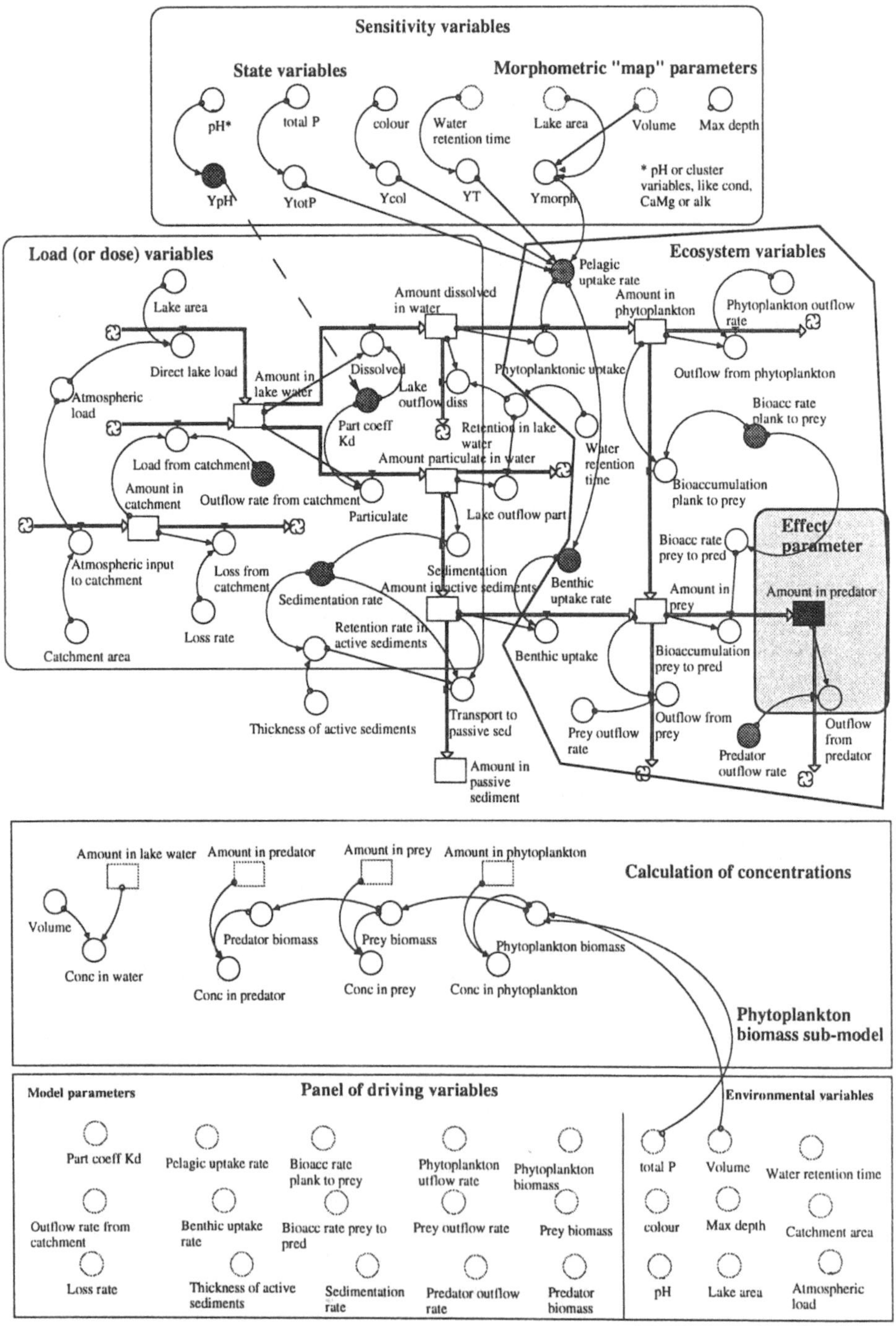

Fig. 9.4. The new mixed model for metals in lakes. (legend continued on following page)

Fig. 9.4. The new mixed model for metals in lakes. *Load (or dose)* parameters are related to the input of metals to the lake (direct load and load from the catchment), the metal amount in the lake water is distributed into dissolved and particulate phases by the partition coefficient (Kd). Sedimentation is net sedimentation per unit of time (the calculation unit is set to 1 year for Hg and 1 month for ^{137}Cs). The *sensitivity* parameters influence biouptake of metals from water to phytoplankton (but they may also be used in other contexts, e.g., to influence the Kd-values, as illustrated by the *dotted line*, or the rate of sedimentation). The biological or *ecosystem variables* include pelagic and benthic uptake, bioaccumulation and retention time in the five compartments (lake water, active sediments, phytoplankton, prey and predator fish). The *effect parameter* is the concentration of the metal in predatory fish (used for human consumption). One panel gives the *calculation of concentrations*, another the *driving parameters* (model variables should, preferably, not be altered for different lakes, while environmental variables must be altered for each lake). The *arrows* between these two panels illustrate the phytoplankton biomass submodel

The model in Fig. 9.4 is based on the following principles:

1. The metal *inflow* to the lake is via direct atmospheric deposition and tributary load from the catchment. The tributary load is regulated by the outflow rate from the catchment. The direct deposition is the net sum of dry and wet fallout (in mass/area units per time units, often as g m^{-2} day^{-1} or bequerel m^{-2} $month^{-1}$) onto the lake surface. The total direct deposition is total fallout times total lake area. The catchment load is a function of amount of X in the catchment (which is a function of fallout and size of the catchment) times an outflow (transfer or retention) rate. We will initially assume that the default value of this rate is 0.0025 (1/year; see Table 9.5), i.e., 0.25% of the amount of X in the catchment area is transported per time unit to the lake. In practice, in natural ecosystems, and for shorter periods of time (<1 year), this rate would depend on many characteristics of the drainage area (inflow and outflow areas, etc.), on precipitation, temperature, evaporation, and the given chemical characteristics of the substance X_1, etc. The rates in dynamic models are sometimes called rate constants, but that gives the wrong impression that rates do not vary in time and space. The rates are variables, just like most of the variables describing the ecosystem (e.g., weight and age of the animals). This means that it is a great challenge for the future to develop predictive models for the most fundamental rates and model variables, like the outflow rate, the rate of sedimentation, and all the biouptake and retention rates.

2. The amount of a metal in the lake is divided into a *particulate* and a *dissolved* phase by means of the *partition coefficient* (Kd; Kd is sometimes called the partitioning or the distribution coefficient), i.e.:

$$\mathrm{Kd} = \frac{\mathrm{X\ in\ solution}}{\mathrm{total\ X}}, \tag{1}$$

$$\text{where total X} = \text{X in solution} + \text{particulate X} \tag{2}$$

There are other ways of defining the partition coefficient, e.g., not as amount in solution divided by total amount in suspension, but as concentration in water relative to concentration on suspended matter. Here, Kd is set to 0.1 as a default value, which means that 10% of X is in solution and 90% is particulate. In practice, it is evident that the Kd-value is *not* a constant, but a variable, which depends on, e.g., (1) the given substance X, (2) lake water pH (and all cluster parameters linked to pH, such as hardness, conductivity, alkalinity, etc.), which influences the equilibrium between X and the particulate phase and the aggregation processes of the carrier particles, (3) the presence of colored substances (humus), which often have a strong affinity to many types of suspended substances (like metals, organic toxins, radioisotopes, etc.), and (4) the character of the particulate phase (if this is clays, humic matter, Fe-oxides/hydroxides, etc.). So, in more extensive ecosystem models, one would need comprehensive submodels to predict the partition coefficient.

3. The *outflow* from the lake is governed by the lake water retention time and the rate of sedimentation. The metals retained in the lake are deposited (=net deposition) on the bottom (regulated by the sedimentation rate), and transported from bioactive to biopassive sediments. These calculations are done on an annual basis for Hg and a monthly basis for ^{137}Cs. The reason for this is that ^{137}Cs has shown to have a rather quick turnover in lake water, plankton and prey, and valuable dynamic (time-dependent) processes would be lost if these calculations were done on an annual basis. Total-Hg seems to have a slower turnover in lake ecosystems. Mercury, unlike ^{137}Cs, appears with a strong organic/lipophilic phase – practically all Hg in predatory fish is methyl-Hg. We'll assume (as default values) that the retention rate for ^{137}Cs in phytoplankton is 1/6 (1/month), for prey 1/12 and for predator 1/24; the corresponding values for Hg are twice as low (i.e., 1/12, 1/24 and 1/48; in 1/month). The biouptake for Cs is, as the empirical data show, rather quick: The perch fry attained high maximum values already during the fall of 1986 (Table 9.1). Sedimentation is net sedimentation during this time period (so resuspension is not accounted for in this model).

4. The metal in solution (dissolved phase) may be picked up by the phytoplankton (regulated by pelagic uptake rate). The *pelagic uptake* rate may be influenced by pH, total-P and color and by the specific lake characteristics (water retention time, area, mean depth, etc.). In traditional mass-balance models, one would multiply an amount (kg) by a rate (1/year) to get a flux. Here, we multiply kg • (1/year) • mod, where mod is a *dimensionless moderator* which describes empirical knowledge how environmental variables (like pH) influence the given flux.

5. The compartment called fish prey consists of, e.g., bottom fauna, zooplankton and small fish. A metal can enter this compartment in *two ways*, from bioaccumulation via phytoplankton (regulated by bioaccumulation rate plankton to prey) and via *benthic uptake* (regulated by benthic uptake rate). Direct uptake of metals from water to prey and predator is not accounted for in this model.

Table 9.5. Default rates and variables in the new mixed model

	Unit	**Hg**	**Cs**
Calculation time		**year**	**month**
Partition coefficient, Kd	1/time	0.1	0.1
Outflow rate from catchment	1/time	0.0025	**function of time** (0.035 to 0.0025)
Loss rate in catchment	1/time	**0.01**	1 / (**30.2**•**12**)
Thickness of active sediments	m	0.05	0.05
Sedimentation rate	1/time	**0.3**	**0.3 / 12**
Pelagic uptake rate	1/time	**3.25**•$\mathbf{10^{-5}}$•YpH	**21**•**3.25**•$\mathbf{10^{-5}}$•YpH
Benthic uptake rate	1/time	1•YpH	1•YpH
Phytoplankton outflow rate	1/time	**1/1**	1 / (**0.5**•12)
Prey outflow rate	1/time	**1/2**	1 / (**1**•12)
Predator outflow rate	1/time	**1/4**	1 / (**2**•12)
Bioaccumulation rate			
plankton to prey	1/time	0.16	**0.5**•0.16
Phytoplanton biomass	kg ww	from model	from model
Prey biomass	kg ww phytoplankton	0.1	0.1
Predator biomass	kg ww phytoplankton	0.01	0.01
pH		6 and 6.7	6 and 6.7
Color	mg Pt/l	(101)	(101)
Total-P	μg/l	(10.5)	(10.5)
Lake area	m^2	0.8 and 0.7	0.7
Max. depth	m	(10)	(10)
Volume	m^3	0.8•4.2	0.7•4.1
Lake water retention time	time	0.89	0.72
Catchment area	m^2	14.7	16.5

ww = weight weight

6. The concentration of a metal in plankton is given by the amount in plankton, mg X, divided by the plankton biomass (kg wet weight). The plankton biomass is predicted by the model given in Table 9.6. This means that, for our "mean" lake, the phytoplankton biomass is about 2000–2500 kg wet weight (Table 9.7). The biomass of the prey is for simplicity assumed to be 10% of the phytoplankton biomass, and the biomass of the predator 10% of the prey. More elaborate approaches to predict biomasses are discussed by Håkanson and Peters (1995).
7. Then, finally, the metal can enter the target compartment, predatory fish (like pike or large perch) via bioaccumulation prey to predator (regulated by bioaccumulation rate prey to predator). This means that this model gives a mean amount or concentration in the predator. More complex models may be used to distribute the metal to various organs of the predator, like liver, kidneys, skin, muscle, etc.

Table 9.6. Selected regressions illustrating the key role of lake total phosphorus in predictive models. Many biological variables that would normally require determinations from extensive and expensive field and laboratory work may be estimated/predicted from one key, abiotic state variable, total-P. Some variables may be predicted with great precision, others with low. (Peters 1986; Håkanson and Peters 1995)

y-value	Equation	Range for TP	r^2-value	n	Units
r^2>0.7					
Chlorophyll	$y = 0.073 \cdot totP^{1.44}$	3– 300	0.96	77	mg/m^3
Max. prim. prod.	$y = 20 \cdot totP - 71$	7– 200	0.95	38	$mg\ C/m^3/day$
Mean prim. prod.	$y = 10 \cdot totP - 79$	7– 200	0.94	38	$mg\ C/m^3/day$
Nanoplankton	$y = 17 \cdot totP^{1.3}$	3– 80	0.93	23	$mg\ ww/m^3$
Phytoplankton	$y = 30 \cdot totP^{1.4}$	3– 80	0.88	27	$mg\ ww/m^3$
Fish yield	$y = 7.1 \cdot totP$	10– 550	0.87	21	$mg\ ww/m^2/a$
Macrozooplankton	$y = 20 \cdot totP^{0.65}$	3– 80	0.86	12	$mg\ ww/m^3$
Zooplankton	$y = 38 \cdot totP^{0.64}$	3– 80	0.86	12	$mg\ ww/m^3$
Bacteria	$y = 0.90 \cdot totP^{0.66}$	3– 60	0.83	12	million/ml
Net plankton	$y = 8.6 \cdot totP^{1.7}$	3– 80	0.82	23	$mg\ ww/m^3$
Fish	$y = 590 \cdot totP^{0.71}$	10– 550	0.75	18	$mg\ ww/m^2$
Crustacean plankton	$y = 5.7 \cdot totP^{0.91}$	3– 300	0.72	49	$mg\ dw/m^3$
Microzooplankton	$y = 17 \cdot totP^{0.71}$	3– 80	0.72	12	$mg\ ww/m^3$
Blue-greens	$y = 43 \cdot totP^{0.98}$	8–1300	0.71	29	$mg\ ww/m^3$
r^2>0.25					
Benthos	$y = 810 \cdot totP^{0.71}$	3– 100	0.48	38	$mg\ ww/m^2$
r^2<0.25					
Transparency	$y = 0.8 \cdot totP^{-0.28}$	3– 400	0.22	87	m

TP=total-P (in mg/m^3). *n* = no. of lakes used in the regression. *ww* = wet weight. *dw* = dry weight

The focus will now be set on the role of (1) the partition coefficient, (2) a changing pH (related to liming), (3) differences in biouptake pattern (pelagic or benthic) and (4) differences in retention (turnover) in the biological compartments (a slower retention for Hg and a faster retention for ^{137}Cs). All data on rates and model variables are given in Table 9.5, and interesting data on amounts and fluxes in Table 9.7.

Table 9.7. Fluxes and amounts for Hg and ^{137}Cs in the mixed model

A. Hg: Fluxes (max. values, not steady-state) to and from the lake, in g Hg/year	
Direct lake load	3.7
Load from catchment	9.6-12.7
Lake outflow from dissolved phase	1.5
Lake outflow from particulate phase	12.5
Sedimentation	2.7
Transport to passive sediments	2.0
B. Hg: Amount in the compartments, max. values in g Hg	
Phytoplankton	0.000212
Prey	0.05
Predator	0.03
Dissolved water phase	1.1
Particulate water phase	9.0
Active sediments	221
Catchment	15,202
C. ^{137}Cs: Fluxes (max. values, not steady-state) to and from the lake, in Bq/month	
Direct lake load	$17.5 \cdot 10^6$
Load from catchment	$8.6 \cdot 10^6$
Lake outflow from dissolved phase	$0.27 \cdot 10^6$
Lake outflow from particulate phase	$2.3 \cdot 10^6$
Sedimentation	$0.5 \cdot 10^6$
Transport to passive sediments	$2.4 \cdot 10^6$
D. ^{137}Cs: Amount in the compartments, max. values in Bq	
Phytoplankton	40,650
Prey	1910
Predator	171
Dissolved water phase	$2.3 \cdot 10^6$
Particulate water phase	$20 \cdot 10^6$
Active sediments	59,150
Catchment	$350 \cdot 10^6$
E. Biomasses, in kg wet weight	
Phytoplankton	2317
Prey	231.7
Predator	23.2

9.6.1 Model Calibration: Dimensionless Moderators

This section presents a method to define dimensionless moderators (Håkanson 1991). Figure 9.5A gives the relationship between scattered data from 817 lakes (data from 1981–86) on lake *pH and* Hg_{pi} (Hg-content in 1-kg pike in mg/kg ww; data from Nilsson et al. 1989). The moderator is used to express the empirical fact that lake acidity (pH) influences uptake of Hg in pike. Instead of building big causal submodels to describe such complicated relationships, the dimensionless moderator accomplishes this in a simple, general manner. So, the basic idea with this calibration (Fig. 9.5) is to use the empirical data between Hg_{pi} and pH from as many lakes as possible to express this highly significant relationship ($p<0.001$; $r^2=0.20$; $n=817$). There are no lakes with a higher mean pH than 8 in Sweden, and *when pH is 8,* one can note from the regression that this corresponds to a Hg_{pi}-*value of about 0.2*, i.e., about the natural, preindustrial reference value. One can also note that for *pH 4 the* Hg_{pi}*-value is about 1.75*. This means that for the entire pH-range of 4 to 8, there is a range in Hg_{pi}-values from 0.2 to 1.75, i.e., *a factor of about 9*. From this empirical knowledge, one can construct a dimensionless moderator (YpH), in this case a *borderline moderator*, because the moderator is set to *1 for pH=8* and *9 for pH=4*. The general formula for a borderline moderator for a given substance X is:

$$YX = 1 + X_{amp} \cdot (X_{emp}/X_{bord} - 1) \tag{3}$$

where
YX = borderline moderator (dimensionless)
X_{amp} = *amplitude* value for X (dimensionless)
X_{emp} = *empirical* value for X in a given lake (specified dimension, e.g., mg Hg/kg ww)
X_{bord} = *borderline* value for X (specified dimension, e.g., mg Hg/kg ww).

If we set X=pH, Fig. 9.5A illustrates the borderline moderator. This is a general technique in which YpH – and other moderators – may be defined, and Fig. 9.5B illustrates the definitions of a similar moderator for ^{137}Cs in predator versus pH36 (mean pH from 36 consecutive months). That moderator is based on empirical data from 41 Swedish lakes for 1988 and 1989 (the data from 1986 and 1987 were omitted in this calibration since the conditions then were far from steady-state). This moderator is similar, but not identical, to the moderator for Hg. The lower pH, the higher the concentrations of elements like Hg and ^{137}Cs in fish.

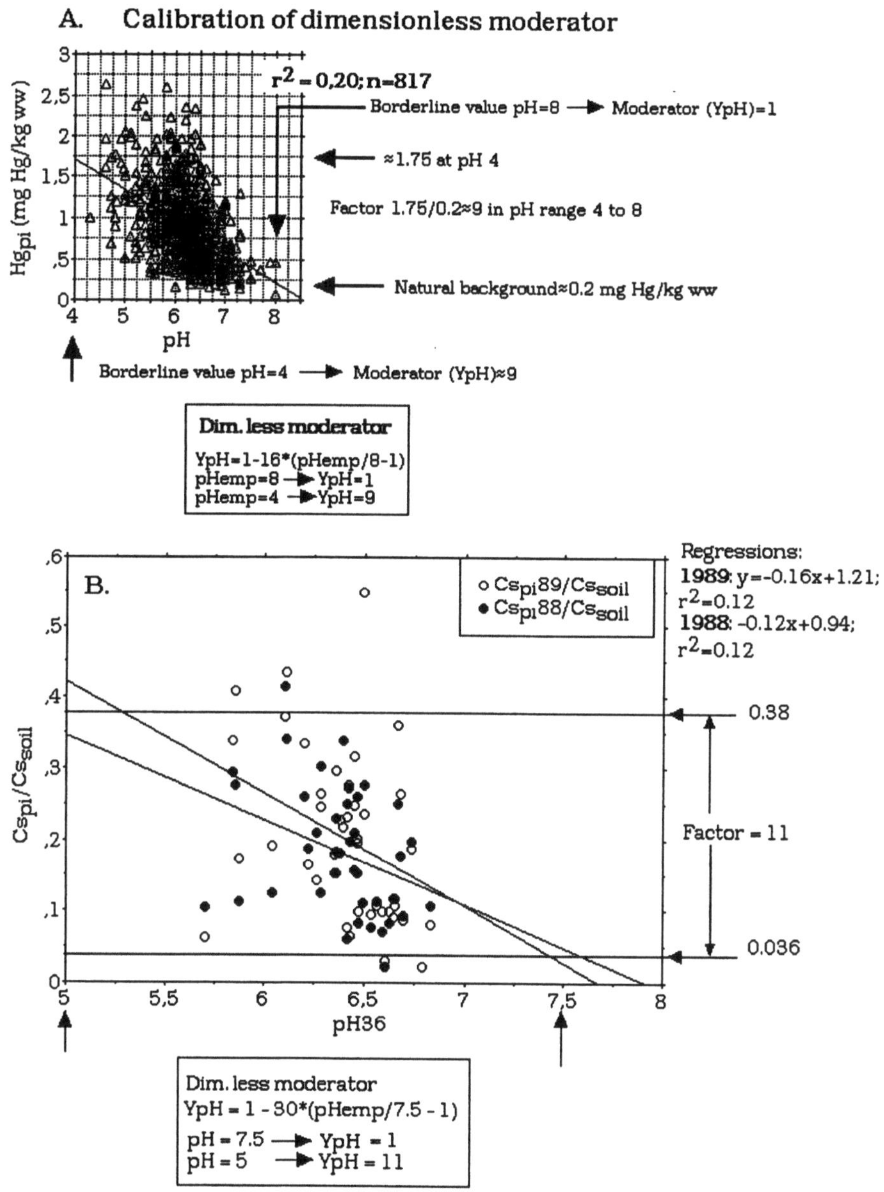

Fig. 9.5A. The relationship between scattered data on lake *pH* and Hg_{pi} (Hg-content in 1-kg pike) from 817 small Swedish lakes from the period 1981–86, and definition of a dimensionless borderline moderator expressing the impact of pH on Hg_{pi}. (Data from Nilsson et al. 1989)
B. Definition of dimensionless moderator (*Dim.less*) for radiocesium in lakes. Empirical data from 41 Swedish lakes (data from Håkanson 1991). Cs_{pi} = ^{137}Cs in pike in Bq/kg ww, Cs_{soil} = fallout in Bw/m^2; *pH36*=mean lake pH from 36 months. *Regression lines* for data for 1988 and 1989

In the following simulations, the other sensitivity factors (color, total-P and morphometry) were kept constant and pH varied to simulate effects of limings on the biouptake of the two test substances, Hg and Cs. It is possible to apply a given dimensionless moderator on many rates and model variables. This is schematically illustrated in Fig. 9.6 by the dotted arrow from YpH to the partition coefficient, since the water chemical conditions (pH, alkalinity, etc.) could influence the way metals are bound to carrier particles.

The result of the calibrations for Hg is shown in Fig. 9.6. Realistic concentrations for a "mean" Swedish lake (given by the data in Table 9.2) were simulated. A liming was conducted year 20 (illustrated by curve 5). pH then increased from 6 to 6.7. These tests are meant to illustrate first the steady-state adjustments and then the recovery processes linked to the liming.

The mean annual value of total-Hg in lake water should be about 5 (ng/l) – it is calibrated to be 4.6 in this modeling setup (Table 9.5); the mean concentration in prey (perch fry) should be 0.18 (mg/kg ww) – this model gives 0.19; the mean

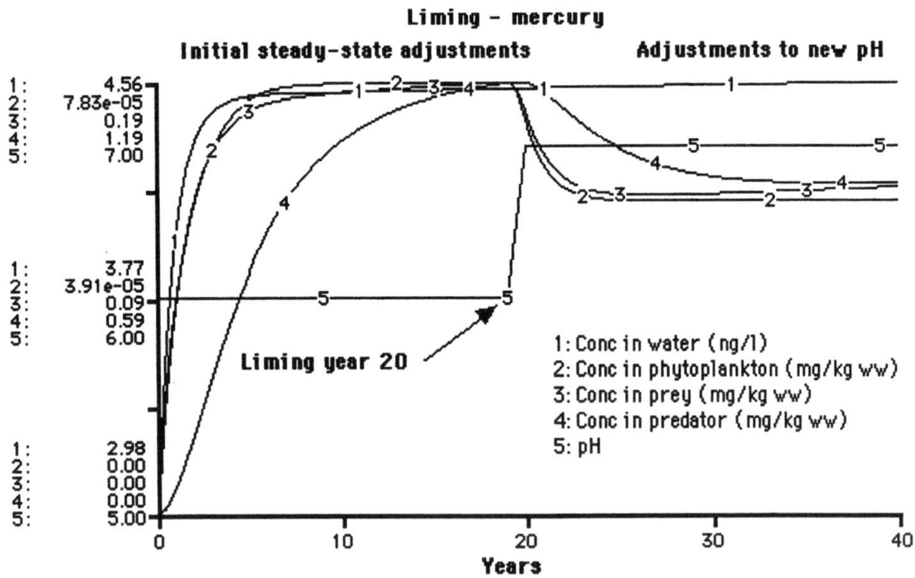

Fig. 9.6. Model simulation for Hg. *Curve 1* gives the values for Hg-concentration in lake water. The values do not reach a perfect steady-state level due to the fact that so much Hg has been accumulated in the catchment area during the last 50 years (initial amount in the catchment is 15 000 g in the entire catchment; situated in central Sweden; data for our "mean" lake), the outflow rate from the catchment is 0.0025 (1/year). *Curve 5* illustrates a simulated liming during year 20 which would increase pH from 6 to 6.7. *Curves 2, 3* and *4* show the effects of such a liming on Hg-concentrations in phytoplankton, prey (bottom fauna, zooplankton, small fish) and predatory fish (pike, large perch, etc.)

concentration of Hg in predator (1-kg pike) should be 1.19 – this model gives 1.19. This applies for lakes with a mean annual pH of 6.0 (≈ the mean value before liming). Figure 9.6 first illustrates how the steady-state concentrations are reached in a theoretical situation where the initial amount of mercury in each compartment is set to zero. It would then take about 20 years to achieve steady-state conditions from such (unrealistic) initial conditions. These simulations are valid for a lake situated in central Sweden were the atmospheric fallout is 10 g Hg/km^2 per year and the total amount of Hg in soils of the catchment is 15 000 g (Lindqvist et al. 1991). One can note the slow recovery of the Hg-concentration in predator (curve 4), and the much faster response for prey (curve 3) and phytoplankton (curve 2). The Hg-concentration in water (curve 1), as expected, is not affected by the liming. The Hg-concentrations predicted by the model as a result of the liming are realistic (like 0.85 for Hg in predator).

9.6.2 Sensitivity Tests

Sensitivity tests have been carried out to identify the most important model variables for calculations of the concentration of Hg and ^{137}Cs in predator fish, and to illustrate how the model works.

To obtain a comparison with the results in Fig. 9.6, pH is also here given by curve 5 in Fig. 9.6. The effects of different Kd-values (0.025 in geometric steps to 0.4) on the Hg-concentration in predator are simulated first. Figure 9.7A shows that the choice of the Kd-value does *not* seem to be very important for the prediction in this particular case (the outcome depends, however, also on the choice of the rates for pelagic and benthic biouptake). Figure 9.7B shows another sensitivity test were the outflow rate from the catchment is varied (0.025 in three steps to 0.00025). One can note a "platform" in the prediction for the two lower outflow rates and that such a high outflow rate as 0.025 (2.5% of the total amount of Hg in the catchment transported to the lake per year) give unrealistically high values of Hg in predator. Figure 9.7C gives the results of a test where the bioaccumulation rate plankton to prey is varied (0.04 in geometric steps to 0.64).

It is evident that high values of this rate will cause a high value of Hg in predator, since this would mean that much Hg is picked up by the prey per unit of time. In this modeling setup, it is assumed (as default value) that the bioaccumulation rate plankton to prey is 0.16 (1/year). Figure 9.7D gives the results of a test where the sedimentation rate is varied (0.05 in geometric steps to 0.8). A sedimentation rate of 0.05 means that 5% of the particulate Hg suspended in the lake is deposited per year. One can note the "platform" for sedimentation rates is less than about 0.4, and that the model is sensitive to the choice of sedimentation rates higher than 0.4 (per year). It has been demonstrated (Håkanson and Jansson 1983) that suspended matter often has a sedimentation rate in the range 0.3 to 0.4 in Swedish lakes. This is, of course, not a constant, but a variable, dependent on the same factors as the partition coefficient, e.g., pH, lake water retention time, etc. In these

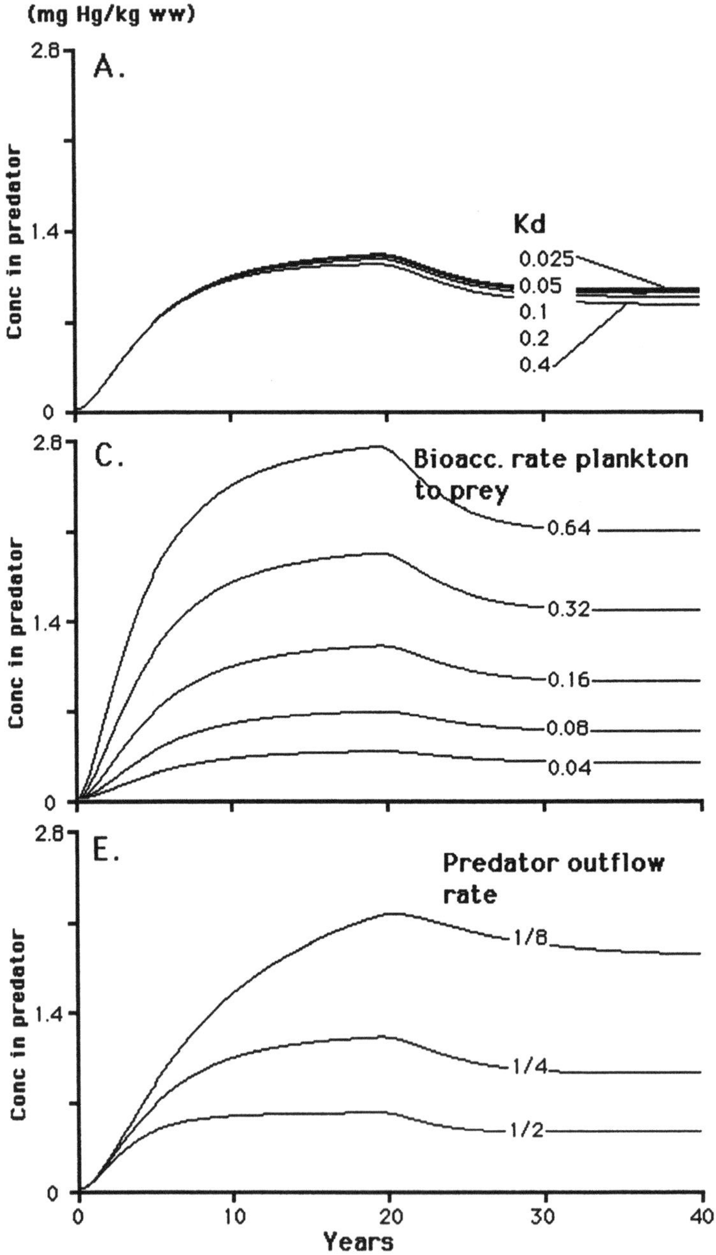

Fig. 9.7. Sensitivity tests for Hg. Simulated effects on Hg-concentrations in predatory fish when one model parameter is varied while all other model parameters are kept constant.
A. Results for different partition coefficients, *Kd*. Default value is 0.1
C. Different *bioaccumulation rates plankton to prey*. Default value is 0.16
E. Different *predator outflow rates*. Default value is 1/4

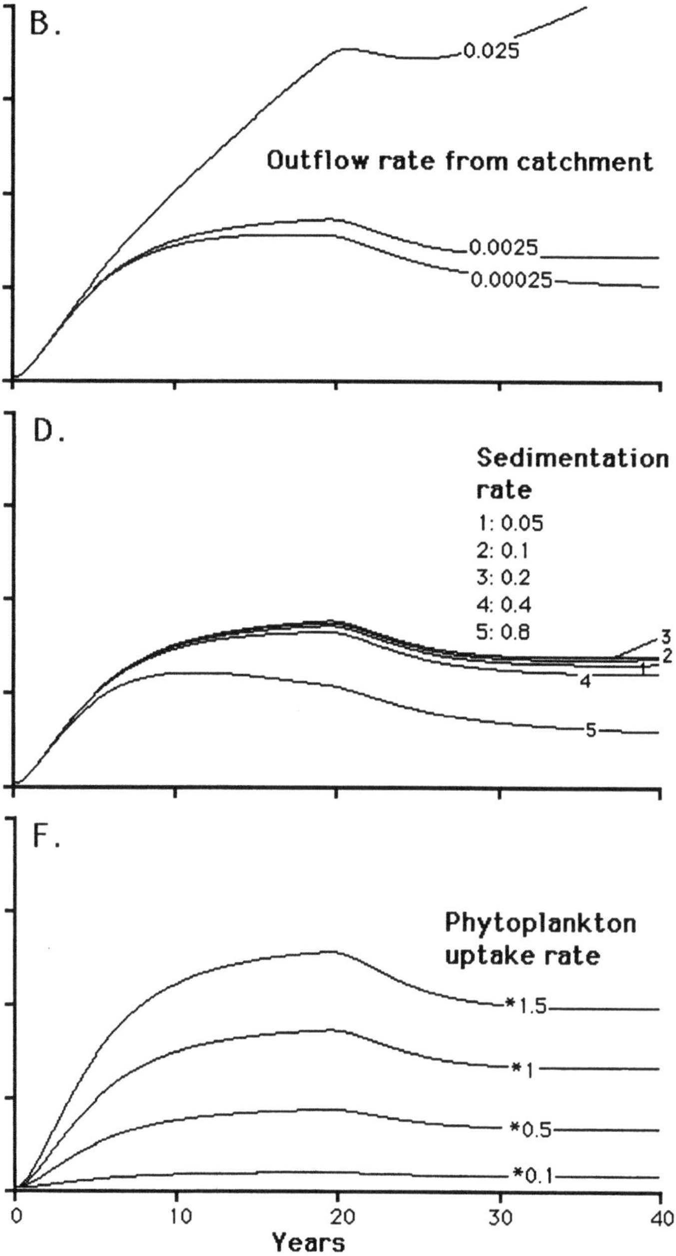

Fig. 9.7. continued
B. Different *outflow rates from the catchment*. Default value is 0.0025
D. Different *sedimentation rates*. Default value is 0.3
F. Different *phytoplankton uptake rates*. Default value is *1

simulations, the sedimentation rate is set to 0.3 (1/year) as a default value. Figure 9.7E gives the results of a test where the predator outflow rate is varied (1/2, 1/4 and 1/8). One can note that the model is very sensitive to the choice of the value of this rate. It has been demonstrated (Meili 1991b) that a retention time of about 4 years ought to be realistic for Hg in predatory fish. Figure 9.7F gives the results of a sensitivity test where the phytoplankton uptake rate is varied. The rate is influenced by the dimensionless moderator for pH (YpH). This rate describes the fraction of the dissolved total-Hg picked up by phytoplankton each year. *1.5 means that the default rate is multiplied with 1.5. It is obvious that high values of the phytoplankton uptake rate will cause high values of Hg in predator, and vice versa.

9.7 Results and Interpretations

From these calibrations and sensitivity tests, one can conclude that the model behaves in a logical manner (as perceived by the modeler, but this may not necessarily be how nature functions). The results of the Liming–mercury–cesium project will now be discussed, focussing on the:

(1) Role of limings; why can a liming cause significant effects on Hg-concentrations in prey fish but no clear effects for ^{137}Cs two years after the remedies,

(2) Relative importance of benthic versus pelagic uptake, and

(3) Role of the partition coefficient.

Figure 9.8 gives the model prediction for ^{137}Cs (with the model parameters from Table 9.5). From Table 9.1 (which gives mean annual data), and from Fig. 9.8A, one can note:

- Average max. annual value of ^{137}Cs in lake water in 1986 (year 1) should be about 2–3 (Bq/l) – the max. *monthly* value according to the model prediction is 4.9, which looks reasonable.
- Average max. annual value of ^{137}Cs in prey (perch fry) in 1986 should be about 8200 (Bq/l) – the max. monthly value according to this model is 8000, a good correspondence
- Average max. annual value of ^{137}Cs in predator (1-kg pike) in 1989 (year 3) should be about 4800(Bq/l) – the max. monthly value according to this modeling setup is 7200. So also in terms of order of magnitude, these results look reasonable. This applies to a lake with a mean pH of 6.0.

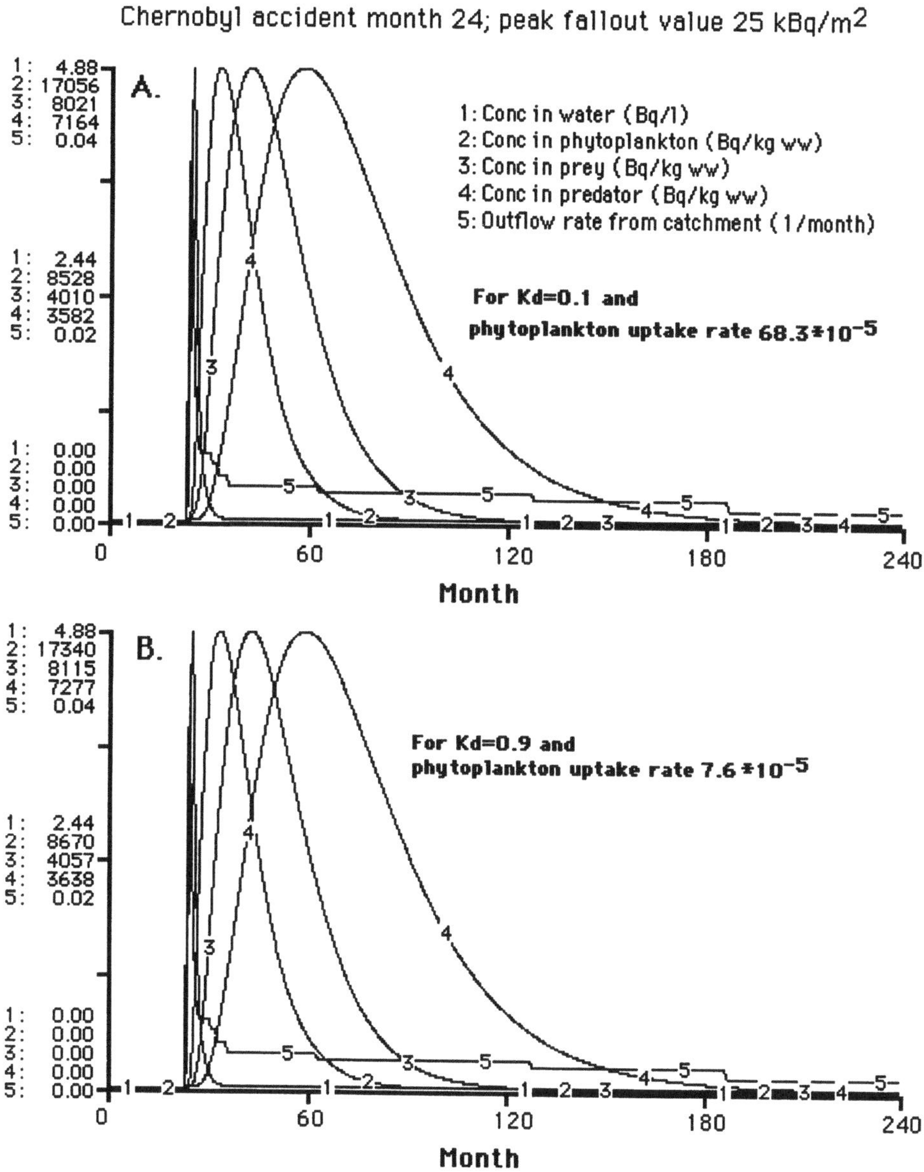

Fig. 9.8. A. Model simulation for ^{137}Cs. The driving function is an atmospheric fallout during month 24, with a peak value of 25 kBq/m^2. *Curves 1, 2, 3* and *4* show the model-predicted concentrations in water, phytoplankton, prey and predatory fish. *Curve 5* gives the assumed time-dependent outflow rate from the catchment. Default Kd=0.1; default phytoplankton uptake rate 68.3 • 10^{-5} (1/month)

B. The same type of model simulation but for Kd=0.9 and phytoplankton uptake rate (68.3/9) • 10^{-5} (1/month)

The time of year of the fallout (in most of the afflicted area) was shortly before the peak of the spring flood, which was unusually massive in large parts of Sweden in 1986. Between 0.5 and 10% of the catchment area's total fallout was probably transported to the lakes during May 1986 (Bergman et al. 1988; Håkanson et al. 1988; Sundblad et al. 1989). During the years 1987–89, the total contribution from the catchment area was probably less than 0.5% of the total deposition. The initial transport to Nordic lakes, and particularly those in northern Sweden, was probably considerably higher than in central European areas with corresponding fallout levels (Santschi et al. 1990). Thus, there exists a pronounced time-dependence for the outflow rate for ^{137}Cs after Chernobyl. A function (curve 5 in Fig. 9.8A) has been applied to describe this. With this function, the monthly rate varies from about 4% (0.04) during the first month after the fallout to less than 0.25% (0.0025) after month 10.

It has been argued that ^{137}Cs has a large particle affinity (Broberg and Andersson 1989; Riise et al. 1990; Salbu et al. 1991), but that may not be generally valid. To emphasize the uncertainty in these matters, tests with different Kd-values have been conducted. Figure 9.8A gives results for Kd=0.1 and Fig. 9.5B for Kd=0.9. One can note that about the same Cs-concentrations in water, phytoplankton, prey and predator can be obtained for the two Kd-values – but only if at the same time the phytoplankton uptake rate is changed by a factor of 9.

Figure 9.9 presents some interesting results. The aim of this test is to compare the consequences of a liming (lake pH is raised in month 36 from 6 to 6.7 and kept at this level till month 96) following a fallout peak (25 units during month 24) for both Hg- and Cs-concentrations in prey (Fig. 9.9A) and predatory fish (Fig. 9.9B). To be able to compare the two substances, the modeling setup given in Table 9.5 has been used for both "Hg" and ^{137}Cs, but it should be noted that the assumptions concerning a zero background and a peak fallout are not realistic for Hg. From Fig. 9.9, one can note that this simulated liming will have a greater effect on "Hg" than on ^{137}Cs in both prey and predator. The liming will, however, also have a positive effects on ^{137}Cs in predator, but not until month 60, i.e., 3 years after the liming. This is, in fact, also what the empirical data suggests. The model could describe this, which is interesting. But how can this be interpreted?

Figure 9.10 tries to answer this question by direct comparisons between "Hg" and "Cs" in water, in phytoplankton (Fig. 9.10B), and in prey and predator (Fig. 9.10A). One can note significant differences. This is due to two factors: The longer outflow rates (twice as long for "Hg") and the faster uptake for "Cs"; the pelagic uptake rate is $21 \cdot 3.25 \cdot 10^{-5}$ (1/month) for "Cs" and $3.25 \cdot 10^{-5}$ (1/year) for "Hg". So, the model and the processes accounted for in the model can describe the empirical results. But it should be stressed that similar results could also be obtained by other combinations of the rates in the model.

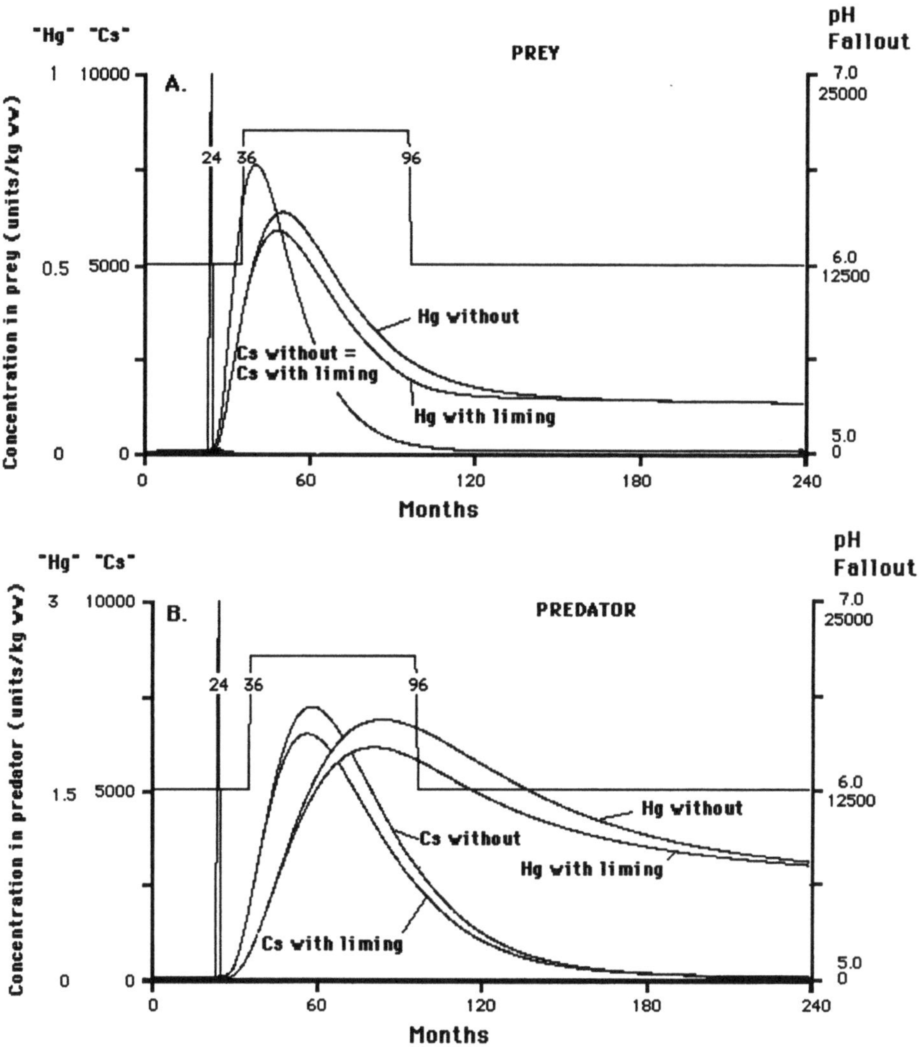

Fig. 9.9. Comparison of model predicted effects of a liming (*curve* marked *36* and *96* for the months when pH was changed from 6 to 6.7 and back to 6) on the concentrations of ^{137}Cs and "Hg" in (**A**) prey and (**B**) predatory fish. These calculations are related to a peak fallout during month 24 (atmospheric load 25 units, kBq/m^2 or g/km^2). The quotation marks have been used because this is a simulation meant to illustrate a hypothetical model comparison between "Cs" and "Hg"

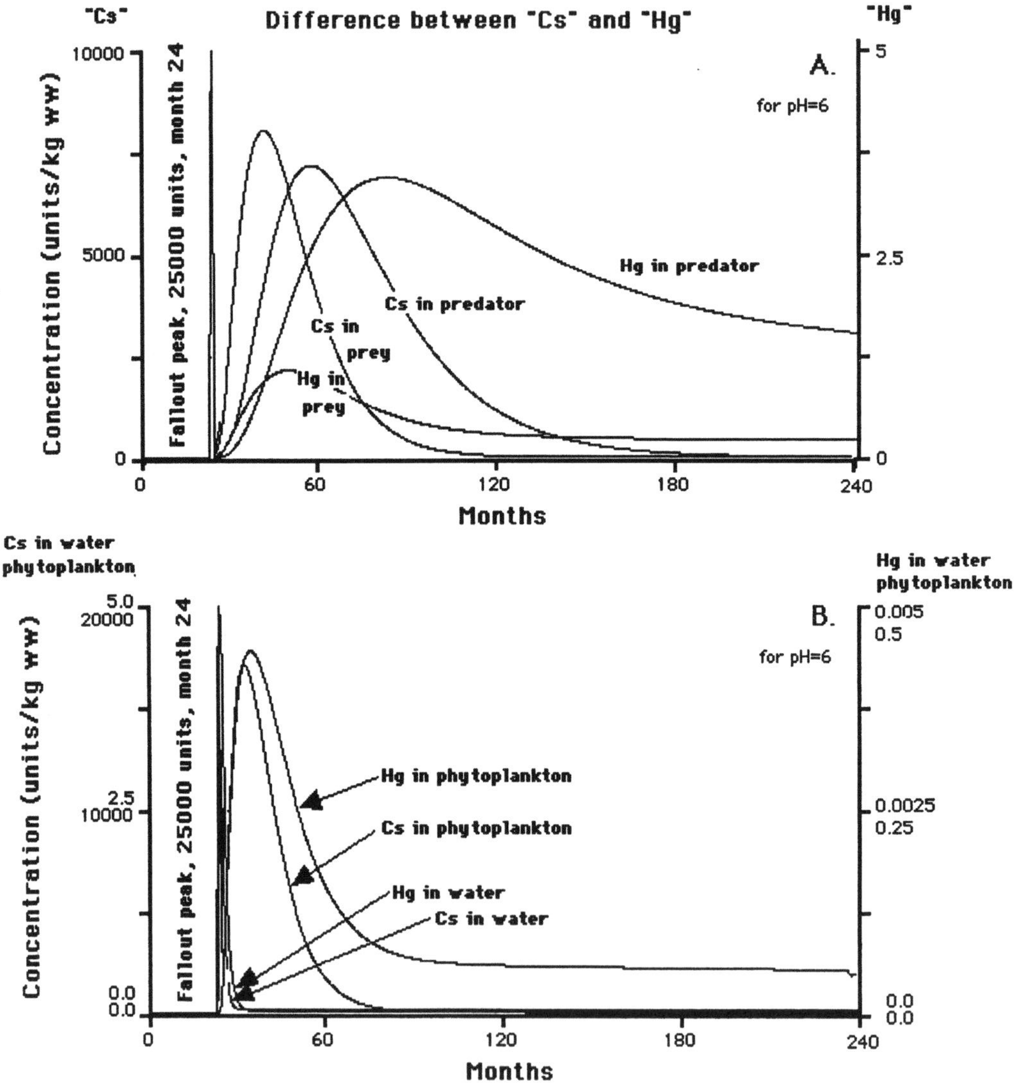

Fig. 9.10. Comparison of model predicted concentrations of "^{137}Cs" and "Hg" in (**A**) predatory fish and in (**B**) prey, water and phytoplankton after an peak fallout during month 24 (atmospheric load = 25 units)

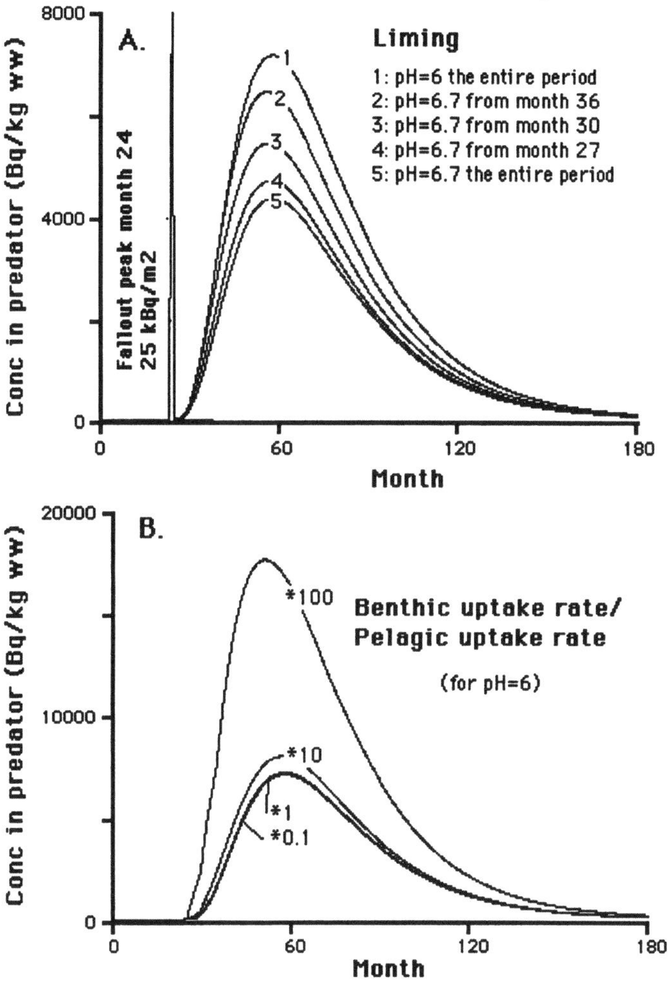

Fig. 9.11. Sensitivity tests for ^{137}Cs. All sensitivity tests are related to a peak fallout of 25 kBq/m^2 during month 24
A. Results for different pH corresponding to different limings (3, 6 and 12 months after the fallout month) when pH has been increased from 6 to 6.7
B. Results for different ratios between benthic and pelagic uptake rate. Default value is *1

Figure 9.11 presents sensitivity tests to give further insights into how the model works. Figure 9.11A and B show that the conditions during fallout are of crucial importance for the biouptake of ^{137}Cs in fish, and that it is inefficient to reduce the Cs-concentration in fish by applying remedies *after* the fallout. In these simulations, this has been tested by increasing pH from 6.0 to 6.7 at month 36 (1 year after the fallout), month 30 and month 27.

Figure 9.11B is a sensitivity test to demonstrate the role of different food

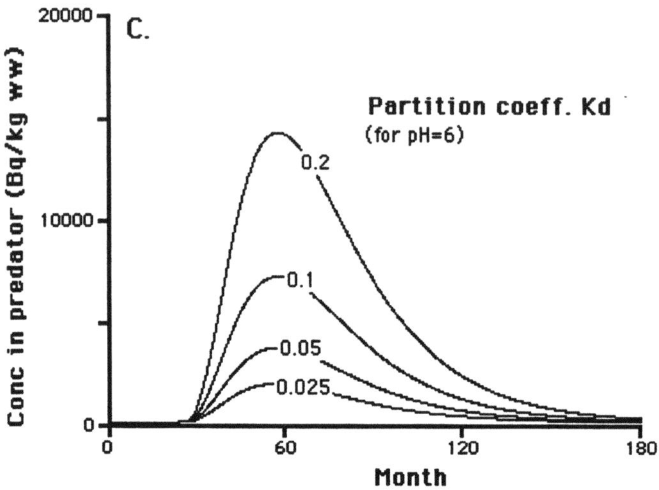

Fig. 9.11. Continued. Sensitivity tests for ^{137}Cs. All sensitivity tests are related to a peak fallout of 25 kBq/m^2 during month 24
C. Results for different values of the partition coefficient (Kd)

choices (different relationships between pelagic and benthic biouptake). The concentration of ^{137}Cs in fish varies strongly with time between different fish species and sizes, mainly depending on the food intake. The benthic flux is generally considered to be most important for the Cs-uptake (Hammar et al. 1989; Andersson et al. 1990; Håkanson 1991; Meili 1991b), whereas pelagic is generally most important for Hg-uptake. One-year old perch mainly live on zooplankton and would, therefore, react faster to environmental changes than the fish-eating pike, which is about 5-years old in this size class (1-kg) in the present lakes. It is noteworthy that the model would predict much higher Cs-concentration in predatory fish if one sets the benthic biouptake 100 times larger than the pelagic uptake. Thus, the model is very sensitive to the choice of the pelagic and benthic biouptake rates.

From Fig. 9.7A, which illustrates the role of the partition coefficient in a sensitivity test for Hg in predator, there is the impression that the model is not very sensitive to the choice of the Kd-value. This is true – but only under certain conditions: if a similar sensitivity test for Cs-concentration in predator is made, setting lake pH to 6, then the model would be very sensitive to alterations in Kd-values, as illustrated in Fig. 9.11C.

Figure 9.12, finally, gives a simulation to illustrate an important general principle for metals in lakes, namely biological dilution. This figure gives calculated results of different values of lake total-P on the effect parameter, which is here Hg-content in top predator. A liming was also simulated year 20, increasing mean annual pH from 6 to 6.7. The model predicts that the biological Hg-uptake will decrease with increased lake eutrophication, which agrees with empirical knowledge (Håkanson 1980).

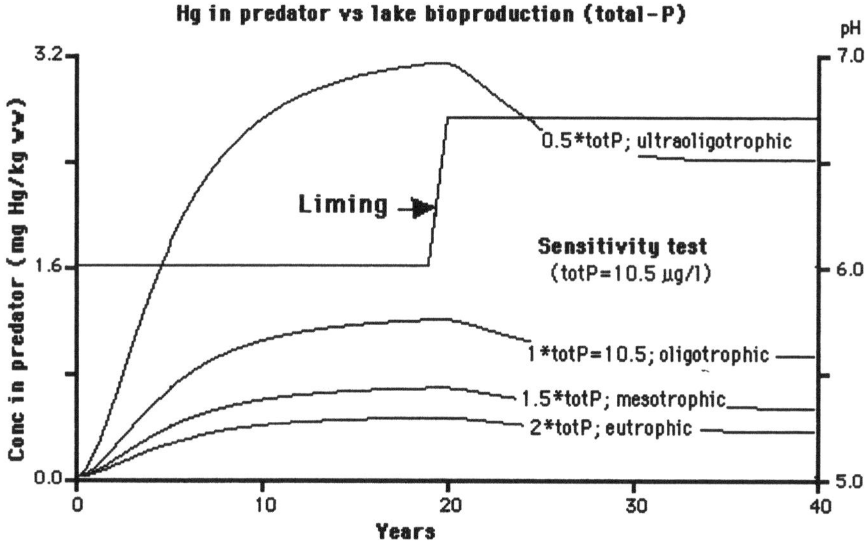

Fig. 9.12. Simulated effects of different lake tot-P (total phosphorus) on the Hg-concentration in top predator. A simulated liming year 20 increases mean annual lake pH from 6 to 6.7. This is one example of "biological dilution", i.e., the biological Hg-uptake decrease with increased lake eutrophication

9.8 Conclusions

There are several possible strategies to reduce metal concentrations in freshwater fish. Great and principally interesting differences exist between metals like Hg, which partly appear in an organic (lipophilic) phase (and not just as Hg^0, Hg^{2+}, etc.) and metals like ^{137}Cs, which only appear in inorganic phases. The amount of metals in fish depends foremost on the biouptake route (pelagic or benthic uptake) and the retention of the metal in fish. The bioavailability can be regulated by a number of chemical and physical variables, such as pH, total-P and color.

This knowledge is clearly of importance to the success or failure of remedial strategy to reduce metal concentrations in fish. Measures that focus on the sediment should, in general for lakes, have poorer chances of success, since the basic problem is not metals in sediment as such, but that the sediment becomes contaminated. Measures intended to eliminate the total amount of metals in lakes are probably also less effective since the pool of metals in the lake water is frequently renewed and since it is the biologically available proportion of this pool that is most interesting. Greater possibilities for a successful result may be expected from measures that reduce the input of metals to the lakes from the catchment area, remedies to reduce the flow via the nutrient chains, e.g., intensive fishing, and methods to reduce the biologically available proportion of the total lake dose, e.g., lake liming and additions of counteractive elements like selenium for mercury and potassium for radiocesium.

The only really satisfactory, long-term alternative in remedial contexts is to minimize the contamination of air, soil and water. All other alternatives (e.g., all the methods discussed in this chapter), regardless of how essential they may be in a short-term perspective, do not influence the basic cause, contamination, but only its manifestations.

The biological turnover of organic (lipophilic) elements like Hg is generally slow; the half-life for Hg in pike is several years. This is important in remedial contexts. A liming may increase a lake's pH from 5 to 6 in a couple of days but the Hg-reduction in fish, that may be linked to this measure, will proceed slowly. If a higher pH is maintained, then this will influence a number of chemical and biological conditions in the lake, and also the Hg uptake in fish. The Hg-concentration in fish will adapt to the new conditions, but this will take a relatively long time.

It should be noted that the new mixed model is intentionally rather simple. There are, e.g., no compartments for algae, periphyton, bacteria and bottom fauna, and no direct uptake from water to fish. Mixed models of this type may use different types of dimensionless moderators and different types of retention techniques (like the traditional retention coefficients used in this model or techniques based on the half-life) to improve the predictive power. Mixed models have many advantages: they can be designed to give specific predictions for just a few y-variables, including the dynamic properties (time dependencies) of traditional dynamic models; they can be built in a logical manner, accounting for empirical knowledge in a simple manner, and be calibrated and validated as other models.

There are significant gaps in our present knowledge concerning the distribution, biouptake, retention and ecological effects of metals in lakes, for example:

- There exist no practically useful, empirically validated load models (effect-dose-sensitivity models) for metals (except for Hg and ^{137}Cs; see Håkanson 1991) in aquatic ecosystems (rivers, lakes and marine areas).

- There are great uncertainties concerning most biouptake and retention rates in dynamic models for metals in most types of aquatic environments.

- Most models for metals in aquatic ecosystems today are derived for specific areas and they would have very low predictive power if applied in other areas (lakes).

- Much holistic work needs to be done to be able to predict effects of metals on ecosystems.

References

Andersson P, Grahn P, Hörnström E, Nyberg P, Dahlquist K (1989) Liming and fertilization in lakes Rammsjön and Ämten, Örebro county 1980–1987. An attempt to reduce the Hg-concentration in fish (in Swedish). SNV Report 35874, 119 pp

Andersson T, Forsgren G, Håkanson L, Malmgren L, Nilsson Å (1990) Radioactive caesium in fish in Swedish lakes after Chernobyl (in Swedish). SSI Rapport 90–04, 41 pp

Andersson T, Håkanson L, Kvarnäs H, Nilsson Å (1991) Remedial measures against high levels of radioactive cesium in Swedish lake fish (in Swedish). SSI Rapport 91–07, 114 pp

Bengtsson Å, Hansson P (1990) Measures against mercury and caesium in lake fish and wild animals. Project 3. Environmental project Sundsvall-Timrå (in Swedish) 29 pp

Bergman R, Danell K, Ericsson A, Grip H, Johansson L, Nelin P, Nylén T (1988) Uptake, re-distribution and transport of radionuclides in a forest ecosystem (in Swedish). FOA-rapport E 40040, Sept. 1988, SLU, Umeå

Björnberg A, Håkanson L, Lundbergh K (1988) A theory on the mechanisms regulating the bioavailability of mercury in natural waters. Environmental Pollution, 48:53–61

Black VS (1957) Excretion and osmoregulation. In: Brown ME (ed) The Physiology of Fishes. Academic Press, New York, 1:163–205

Boudou A, Ribeyre F (1990) Aquatic ecotoxicology: Fundamental concepts and methodologies. CRC Press, Boca Raton

Broberg A, Andersson E (1989) Circulation of caesium in limnic ecosystems (in Swedish). Inst. of Limnology, Uppsala Univ, 30 pp

Carlsson S (1978) A model for the turnover of ^{137}Cs and potassium in pike (*Esox Lucius*). Health Phys, 35:549–554

Fleishman DG (1963) Accumulation of artificial radionuclides in freshwater fish. In: Klechkovskii VM, Polikarpov GG, Aleksakhin RM (eds) Radioecology. John Wiley, New York, pp 347–370

Förstner U, Müller G (1974) Schwermetalle in Flüssen und Seen. Springer, Berlin Heidelberg New York, 225 pp

Förstner U, Wittmann GTW (1981) Metal pollution in the aquatic environment. Springer, Berlin Heidelberg New York, 486 pp

Göthberg A (1983) Intensive fishing – a way to reduce the mercury level in fish. Ambio, 12:259–261

Hammar J, Notter M, Neumann G (1989) Caesium in char lakes (in Swedish). Report to SSI (project P 378.86 and P 378.88)

Håkanson L (1980) The quantitative impact of pH, bioproduction and Hg-contamination on the Hg-content of fish (pike). Environmental Pollution, B, 1:285–304

Håkanson L (1984) Aquatic contamination and ecological risk. An attempt to a conceptual framework. Water Res, 18:1107–1118

Håkanson L (1990) An operative system for environmental consequence analysis for aquatic ecosystems. - In: Baudo R, Giesy JP, Muntau H (eds) Sediments: chemistry and toxicity of in-place pollutants, Lewis Publishers, Michigan, pp 365–390

Håkanson L (1991) Ecometric and dynamic modelling – exemplified by caesium in lakes after Chernobyl. Springer, Berlin Heidelberg New York, 162 pp

Håkanson L, Andersson P, Andersson T, Bengtsson Å, Grahn P, Johansson J-Å, Jönsson C-P, Kvarnäs H, Lindgren G, Nilsson Å (1990a) Measures against high levels of mercury in lake fish. Final report from the liming–mercury–cesium projekt (in Swedish with English summary). SNV (Swedish Environmental Protection Agency) Rapport 3818, 189 pp

Håkanson L, Andersson T (1992) Remedial measures against radioactive caesium in Swedish lake fish after Chernobyl. Aquatic Sci, 54:141–164

Håkanson L, Andersson T, Neumann G, Nilsson Å, Notter M (1988) Cesium-137 in perch in lakes from northern Sweden after Chernobyl – present situation, relationships, trends. Swed. Env. Prot. Agency, Report 34 97, 136pp
Håkanson L, Borg H, Uhrberg R (1990b) Reliability of analyses of Hg, Fe, Ca, K, P, pH, alkalinity, conductivity, hardness and colour from lakes. Int Rev ges Hydrobiol, 75: 79–94
Håkanson L, Peters RH (1995) Predictive limnology – methods for predictive modelling. SPB Academic Publishing, Amsterdam, 464pp
Jansson M, Hayman U, Forsberg C (1981) Acid lakes and "biological buffering" (in Swedish). Vatten, 37: 241–251
Johnels AG, Westermark T, Berg W, Persson PI, Sjöstrand B (1967) Pike (*Esox lucius L.*) and some other aquatic organisms in Sweden as indicators of mercury contamination in the environment. Oikos, 18:323–333
Lindqvist O, Johansson K, Aastrup M, Andersson A, Bringmark L, Hovsenius G, Håkanson L, Iverfeldt Å, Meili M, Timm B (1991) Mercury in the Swedish environment. Water Air Soil Pollut, vol. 55, 261 pp
Meili M (1991a) In situ assessment of trophic levels and transfer rates in aquatic food webs using chronic (Hg) and pulsed (Chernobyl ^{137}Cs) environmental contaminants. Verh. Int. Verein. Limnol, 24
Meili M (1991b) Mercury in boreal forest lake ecosystems. Acta Univ. Upsaliensis 336. Thesis, Uppsala Univ
Moberg L (ed)(1991) The Chernobyl fallout in Sweden. The Swed. Rad. Prot. Inst, Stockholm, 633 pp
Monitor (1986) Acid and acidified waters (in Swedish). SNV, Solna, 180 pp
Nilsson Å, Andersson T, Håkanson L, Andersson A (1989) Mercury in lake fish – linkages to selenium and mercury in the mor layer and to historical emissions. SNV Report 3593, 117 pp
Paulsson K, Lundbergh K (1989) The selenium method for treatment of lakes for elevated levels of mercury in fish. Sci Tot Env, 87/88:496–507
Peters RH (1986) The role of prediction in limnology. Limnol Oceanogr, 31:1143–1159
Peters RH (1991) A critique for ecology. Cambridge Univ. Press, Cambridge, 366 pp
Riise G, Björnstad HE, Oughton DH, Salbu B (1990) A study on radionuclide associations with soil components using sequential extraction procedure. J Radioanal Nucl Chem, 142:531–538
Rudd JWM, Turner MA (1983) Suppression of mercury and selenium bioaccumulation by suspended and bottom sediments. Can J Fish Aquat Sci, 40:2218–2227
Turner MA, Furutani A, Swick A, Townsend BE (1983) The English-Wabigoon river system: a synthesis of recent research with a view towards mercury amelioration. Can J Fish Aquat Sci, 40:2206–2217
Salbu B, Björnstad HE, Brittain JE (1991) Fractionation of Cs-isotopes and 90-Sr in snow melt, runoff and lake water from contaminated area in Norway. J Radioanal Nucl Chem
Salomons W, Förstner U (1984) Metals in the Hydrocycle. Springer, Berlin Heidelberg New York, 349 pp
Santschi PH, Bollhalder S, Zingg S, Luck A, Farrenkothen K (1990) The self-cleaning capacity of surface water after radioactive fallout. Evidence from European water after Chernobyl, 1986–1988. Environ Sci Technol, 24:519–527
Sundblad B, Evans S, Bergström U (1989) The turnover of Chernobyl fallout within two catchment areas – Hillesjön and Slägsjön – in the Gävle area, Sweden. Studsvik/NP-89/51
Vernet J-P (ed)(1991) Heavy metals in the environment. Elsevier, Amsterdam, 405 pp

Appendix

Model, Equations, Assumptions and Presuppositions

Compartment: Amount dissolved in water

Amount dissolved in water(t) = Amount dissolved in water(t - dt) + (Dissolved - Phytoplanktonic uptake - Lake outflow diss) • dt
INIT Amount dissolved in water = 1
DOCUMENT: Values in g for Hg and Bq for ^{137}Cs. Default value = 0.
INFLOWS:
Dissolved = Part coeff Kd • Amount in lake water
OUTFLOWS:
Phytoplanktonic uptake = 1 • Pelagic uptake rate • Amount dissolved in water
DOCUMENT: Units are 1/year for Hg, 1/month for ^{137}Cs and 1/unit of time more generally.
Lake outflow diss = Retention in lake water • Amount dissolved in water

Compartment: Amount in active sediments

Amount in active sediments(t) = Amount in active sediments(t - dt) + (Sedimentation - Transport to passive sed - Benthic uptake) • dt
INIT Amount in active sediments = 200
DOCUMENT: Default value 200 g.
INFLOWS:
Sedimentation = Amount particulate in water • Sedimentation rate
OUTFLOWS:
Transport to passive sed = Sedimentation rate • Amount in active sediments • Retention rate in active sediments
Benthic uptake = Benthic uptake rate • Amount in active sediments

Compartment: Amount in catchment

Amount in catchment(t) = Amount in catchment(t - dt) + (Atmospheric input to catchment - Loss from catchment) • dt
INIT Amount in catchment = 15000
DOCUMENT: Default value set to 15000 g Hg for central Sweden, and 30000 g for south central Sweden (Lindqvist et al., 1991). The default value for ^{137}Cs is 0.
INFLOWS:
Atmospheric input to catchment = Atmospheric load • Catchment area • 10^{-6}
OUTFLOWS:
Loss from catchment = Amount in catchment • Loss rate

Compartment: Amount in lake water

Amount in lake water(t) = Amount in lake water (t - dt) + (Load from catchment + Direct lake load - Particulate - Dissolved) • dt
INIT Amount in lake water = 10
DOCUMENT: Default value 10 g.
INFLOWS:
Load from catchment = Amount in catchment • Outflow rate from catchment • (1/3)
DOCUMENT: In g/year.
Direct lake load = Atmospheric load • (1/3) • Lake area • 10^{-6}
DOCUMENT: In g/year.
OUTFLOWS:
Particulate = (1-Part coeff Kd) • Amount in lake water
Dissolved = Part coeff Kd • Amount in lake water

Compartment: Amount in passive sediment

Amount in passive sediment(t) = Amount in passive sediment(t - dt) + (Transport to passive sed) • dt
INIT Amount in passive sediment = 0
DOCUMENT: Default value 0 g.
INFLOWS:
Transport to passive sed = Sedimentation rate • Amount in active sediments • Retention rate in active sediments

Compartment: Amount in phytoplankton

Amount in phytoplankton(t) = Amount in phytoplankton (t - dt) + (Phytoplanktonic uptake + Bioaccumulation plank to prey - Outflow from phytoplankton) • dt
INIT Amount in phytoplankton = 0
DOCUMENT: Default value 0 g.
INFLOWS:
Phytoplanktonic uptake = 1 • Pelagic uptake rate • Amount dissolved in water
DOCUMENT: Units are 1/year for Hg, 1/month for ^{137}Cs.
Bioaccumulation plank to prey = Amount in phytoplankton • Bioacc rate plank to prey
OUTFLOWS:
Outflow from phytoplankton = Amount in phytoplankton • Phytoplankton outflow rate

Compartment: Amount in predator

Amount in predator(t) = Amount in predator(t - dt) + (Bioaccumulation prey to pred - Outflow from predator) • dt
INIT Amount in predator = 0

DOCUMENT: Values in g.
INFLOWS:
Bioaccumulation prey to pred = Bioacc rate prey to pred • Amount in prey
OUTFLOWS:
Outflow from predator = Predator outflow rate • Amount in predator

Compartment: Amount in prey

Amount in prey(t) = Amount in prey (t - dt) + (Benthic uptake - Bioaccumulation prey to pred - Outflow from prey - Bioaccumulation plank to prey) • dt
INIT Amount in prey = 0
DOCUMENT: Values in g.
INFLOWS:
Benthic uptake = Benthic uptake rate • Amount in active sediments
OUTFLOWS:
Bioaccumulation prey to pred = Bioacc rate prey to pred • Amount in prey
Outflow from prey = Prey outflow rate • Amount in prey
Bioaccumulation plank to prey = Amount in phytoplankton • Bioacc rate plank to prey

Compartment: Amount particulate in water

Amount particulate in water(t) = Amount particulate in water(t - dt) + (Particulate - Lake outflow part - Sedimentation) • dt
INIT Amount particulate in water = 9
DOCUMENT: Default value is 0 g.
INFLOWS:
Particulate = (1-Part coeff Kd) • Amount in lake water
OUTFLOWS:
Lake outflow part = Amount particulate in water • Retention in lake water
Sedimentation = Amount particulate in water • Sedimentation rate
Benthic uptake rate = Pelagic uptake rate
DOCUMENT: In 1/year. Default value is set equal to the pelagic uptake rate.
Bioacc rate plank to prey = 0.16
DOCUMENT: It is assumed that 16% (rate 0.16) of the metal amount (for both Hg and ^{137}Cs) in plankton is transported to prey fish per unit of time.
Bioacc rate prey to pred = Bioacc rate plank to prey
DOCUMENT: As a default assumption, the prey-to-predator bioaccumulation rate is set equal to the plankton-to-prey bioaccumulation rate.
Catchment area = $15.6 \cdot 10^6$
DOCUMENT: In m^2.
colour = 101
DOCUMENT: Effects of colour on metal uptake are not simulated in this exercise, but high values of lake colour would generally lower the biouptake of metals. In mg Pt/l.

Conc in phytoplankton = 1000 • Amount in phytoplankton/Phytoplankton biomass
DOCUMENT: Values in mg/kg ww.
Conc in predator = 1000 • Amount in predator/Predator biomass
DOCUMENT: Values in mg/kg ww.
Conc in prey = 1000 • Amount in prey/Prey biomass
DOCUMENT: Values in mg/kg ww.
Conc in water = 10^6 • Amount in lake water/Volume
DOCUMENT: Values in ng/l; g/m^3 = mg/l; or kBq/m^3 = Bq/l.
Lake area = 0.8 • 10^6
DOCUMENT: In m^2.
Loss rate = 0.01
DOCUMENT: It is assumed that 1% of the amount in the catchment may be lost per time unit due to evaporation, chemical transformations, transport to "geological" compartments, etc. The rate of loss for ^{137}Cs is defined from the physical half-life (1/30.2 years).
Max depth = 10
Outflow rate from catchment = 0.0025
DOCUMENT: In 1/year. This rate gives the amount transported per unit of time from the catchment to the lake (a runoff rate). The default value is 0.0025 (i.e., 0.25%) for Hg, and a time-dependent function is used for ^{137}Cs after Chernobyl.
Part coeff Kd = 0.1
DOCUMENT:
Kd = Diss-Me/Tot-Me = 0.1/1 = 0.1.
Pelagic uptake rate = 0.0000325 • YpH • YtotP • Ycol • YT • Ymorph
DOCUMENT: It is assumed that a small fraction of the dissolved element in the lake is picked up by phytoplankton per time unit, i.e., the pelagic uptake rate is small. The benthic uptake rate for Hg and ^{137}Cs is, as a default assumption, set to be equal to the pelagic uptake rate.
Phytoplankton biomass = 10^{-6} • Volume • 30 • total $P^{1.4}$
DOCUMENT: Values in kg wet weight predicted from equation in Table 9.6.
Phytoplankton outflow rate = 1
DOCUMENT: This rate is set to 1.
Predator biomass = Prey biomass/10
DOCUMENT: In kg wet weight.
Predator outflow rate = 1/4
DOCUMENT: If the retention time of the element in predator fish is x years, it is assumed that the retention rate is 1/x; x is generally set to 4 years for Hg and to 2 years for ^{137}Cs.
Prey biomass = Phytoplankton biomass/10
DOCUMENT: In kg wet weight.
Prey outflow rate = 1/2
DOCUMENT: The retention rate is set to 1/2 years for Hg and to 1/1 year for ^{137}Cs.

Retention in lake water = 1/Water retention time
DOCUMENT: In 1/years.
Retention rate in active sediments =
1/((0.3 • Thickness of active sediments)/(0.0015 • Sedimentation rate))
DOCUMENT: If the thickness of the active sediment layer is set to 5 cm (0.05 m), the rate of sedimentation is set to 0.3 (1/year), and if this corresponds to a sediment accumulation of 0.15 cm/year (Håkanson and Jansson, 1983), then the mean age of the active sediment layer is 5 • 0.015 • 0.3/0.3 = 33 years. The corresponding retention rate is 1/33.
Sedimentation rate = 0.3
DOCUMENT: It is assumed that 30% (sed. rate = 0.3; in 1/year) of the particulate phase is deposited per year.
Thickness of active sediments = 0.05
DOCUMENT: The thickness of the active layer is set to 5 cm.
total P = 10.5 • 1
DOCUMENT: Values in μg/l=mg/m^3. Annual values are used for Hg , and monthly values for Cs. Total-P is also used to predict the phytoplankton biomass.
Volume = 4.2 • 0.8 • 10^6
DOCUMENT: Lake volume in m^3.
Water retention time = 0.72
DOCUMENT: Values in years.
Ycol = 1
Ymorph = 1
DOCUMENT: Effects of different lake morphometry on metal uptake will not be simulated in this modelling exercise.
YpH = 1-16 • (pH/8-1)
DOCUMENT: See Fig. 9.5 for definition.
YT = 1
YtotP = 1
Atmospheric load = GRAPH(TIME)
(0.00, 10.0), (1.00, 10.0), (2.00, 10.0), (3.00, 10.0), (4.00, 10.0), (5.00, 10.0), (6.00, 10.0), (7.00, 10.0), (8.00, 10.0), (9.00, 10.0), (10.0, 10.0), (11.0, 10.0), (12.0, 10.0), (13.0, 10.0), (14.0, 10.0), (15.0, 10.0), (16.0, 10.0), (17.0, 10.0), (18.0, 10.0), (19.0, 10.0), (20.0, 10.0), (21.0, 10.0), (22.0, 10.0), (23.0, 10.0), (24.0, 10.0), (25.0, 10.0), (26.0, 10.0), (27.0, 10.0), (28.0, 10.0), (29.0, 10.0), (30.0, 10.0), (31.0, 10.0), (32.0, 10.0), (33.0, 10.0), (34.0, 10.0), (35.0, 10.0), (36.0, 10.0), (37.0, 10.0), (38.0, 10.0), (39.0, 10.0), (40.0, 10.0)
DOCUMENT: In g/m^2 • year. The default is 10 g Hg/km^2 • year for central Sweden and 20 g for south central Sweden (Lindqvist et al. 1991).
pH = GRAPH(TIME)
(0.00, 6.00), (1.00, 6.00), (2.00, 6.00), (3.00, 6.00), (4.00, 6.00), (5.00, 6.00), (6.00, 6.00), (7.00, 6.00), (8.00, 6.00), (9.00, 6.00), (10.0, 6.00), (11.0, 6.00), (12.0, 6.00), (13.0, 6.00), (14.0, 6.00), (15.0, 6.00), (16.0, 6.00), (17.0, 6.00), (18.0, 6.00), (19.0, 6.00), (20.0, 6.70), (21.0, 6.70), (22.0, 6.70), (23.0, 6.70),

(24.0, 6.70), (25.0, 6.70), (26.0, 6.70), (27.0, 6.70), (28.0, 6.70), (29.0, 6.70), (30.0, 6.70), (31.0, 6.70), (32.0, 6.70), (33.0, 6.70), (34.0, 6.70), (35.0, 6.70), (36.0, 6.70), (37.0, 6.70), (38.0, 6.70), (39.0, 6.70), (40.0, 6.70)
DOCUMENT: This is time-dependent pH. When effects of liming are simulated, default values are initially set to pH = 6, and after liming to pH = 6.7.

Part 3

Waste Treatment and Landfill Disposal

10 Waste Treatment: Geochemical Engineering Approach

U. Förstner

10.1 Introduction

Contaminated solid materials, which may affect groundwater pollution include municipal solid wastes, sewage sludge, dredged sediments, industrial by-products, wastes from mining and smelting operations, filter residues from waste water treatment and atmospheric emission control, ashes and slags from burning of coal and oil, and from incineration of municipal refuse and sewage sludge. The problem of "contaminated land" was introduced with the accidential detection of large-scale pollution from industrial waste deposits which had been handled improperly.

Table 10.1. Quantities of solid waste materials (Förstner 1993, after Baccini and Brunner 1991; Neumann-Malkau 1991)

Waste type	kg/capita/year	
Municipal solid waste	130	(Bangladesh)
	520	(U.S.A.)
Sewage sludge (95% Water)	720	(F.R.G.)
Construction waste (incl. soil movements		
– ca. 80% – and road construction waste)	2.200	(F.R.G.)
Industrial and hazardous waste	80	(F.R.G.)
	million tonnes/year	
Dredged materials	580	(World)
Incineration residues	300	(World)
Mine waste + tailings	17,800	(World)
(Natural erosion:	26,700)	

As an example, Table 10.1 lists quantitative figures of the production of waste materials. Municipal solid waste amounts to approximately half a tonne per person per year. Sewage sludge is even more. Construction waste including soil movements is more than two tonnes per person and year, with large differences in quality – most of them unproblematic, a smaller portion, however, highly contaminated. Table 10.1 also lists shredder and tire waste, which is a relative small portion compared with the previous items. However, one has to consider that the manufacture of a lower-middle-class car generates approx. 15 tonnes of wastes, that is 25 times the weight of the car. A considerable proportion of such wastes is related to mining activities, and the quantities of mine tailings are in the same order of magnitude as the actual sediment discharge to oceans.

10.2 Critical Sites Related to Dispersion of Contaminants

With regard to loss of groundwater resources, ascribable to contamination, a United States Library of Congress Report from 1980 lists 1360 well closings in a 30-years span, which can be broken down as follows (Anonymous 1980): metal contamination: 619 wells; organic chemical contamination: 242, including 170 from trichloroethylene used to emulsify septic tank grease; pesticide contamination: about 200 wells; industrial (not defined) sites: 185 wells; leachate from municipal solid waste landfills: 64 wells; and 26 wells affected by high concentrations of chlorides, 23 wells polluted by nitrates. Altogether, there is a significant effect of heavy-metal pollution – about 40% in this study – on the loss of groundwater resources.

Only a few industries or activities account for most of the difficulties with metal problems; these are in particular metal plating, chemical, mining and smelting, battery recycling, wood treating, and oil and solvent recycling (Wilmoth et al. 1991). Not surprisingly, in the United States national priority list of 1986 (Anonymous 1986a) developed by the EPA, landfills account for a high percentage (40%) as chemical wastes were often dumped in municipal landfills.

Comparison of inorganic groundwater constituents upstream and downstream of 33 waste disposal sites in West Germany (Arneth et al. 1989) indicates typical differences in pollutant mobilities, which may partly be related to releases during the acidic phase of the landfill development. High contamination factors (contaminated mean/uncontaminated mean of downstream/upstream groundwater samples from landfills) have been found for boron (>63), ammonia (62) and arsenic (34; this element may pose particular problems during initial phases of landfill operations; Blakey 1984). Heavy metals such as cadmium (CF >6.5), chromium (>5), lead (5.0), copper (4.7) and nickel (3.0) are significantly enriched in the leachates as well.

10.3 Critical Processes for Metal Mobilization from Wastes

Regarding the potential release of metals from solid waste materials, changes of pH and redox conditions are of prime importance. It can be expected that changes from reducing to oxidizing conditions, which involve transformations of sulphides and a shift to more acidic conditions, increase the mobility of typical "B-" or "chalcophilic" elements, such as Hg, Zn, Pb, Cu and Cd. On the other hand, the mobility is characteristically lowered for Mn and Fe under oxidizing conditions. Elements able to form anionic species, such as S, As, Se, Cr and Mo are solubilized, for example, from fly ash sluicing/ponding systems at neutral to alkaline pH conditions (Dreesen et al. 1977; Turner et al. 1982)

The major process affecting the lowering of pH-values (down to pH 2 to 3) is the exposure of pyrite (FeS_2) and of other sulphide minerals to atmospheric oxygen and moisture, whereby the sulphidic component is oxidized to sulphate and the acidity (H^+-ions) is generated. Bacterial action can assist the oxidation of Fe^{2+}(aq) in the presence of dissolved oxygen.

Acidity is perhaps the most serious long-term threat from metal-bearing wastes. Water seeping from mine refuse has been passing increased metal concentrations into receiving waters for decades. The threat is especially great in waters with little buffer capacity, i.e. in carbonate-poor areas, where dissolved-metal pollution can be spread over great distances. The acidity production can develop many years after disposal, when the neutralizing or buffering capacity in a pyrite-containing waste is exceeded. High concentration factors have been determined in inland waters affected by acidic mine effluents.

Primary emissions of high metal concentration occur from waste rocks and tailings and secondary effects on groundwater take place from the ponds. An important and long-term source of metals are the sediments reworked from the floodplain, mainly by repeated oxidation and reduction processes.

The acid-producing potential not only is related to the oxidation of sulphides, but oxidation of organic matter must be considered as well. It has been shown by Swift (1977) that the contribution of protons from organic-N and organic-S in a sample containing approximately 5% organic carbon is equivalent to the acid-producing potential of 1% FeS_2. Studies of the long-term evolution and diagenesis in a sewage sludge landfill and similar natural sediments (peat, organic soils) by Lichtensteiger et al. (1987) suggest that the transformation of organic material will last for geological time scales (10^3 to 10^7 years).

Pore-water data from dredged material from Hamburg Harbour indicate typical differences in the kinetics of proton release from organic and sulphidic sources (Table 10.2). Recent deposits are characterized by low concentrations of nitrate, cadmium and zinc. When these low-buffered sediments are oxidized during a time period of a few months to years, the concentrations of ammonia and iron in the pore water typically decrease, whereas those of cadmium and zinc increase (with the result that these metals are easily transferred into agricultural crops!).

Table 10.2. Metal mobilization from dredged material after land deposition (Maaß and Miehlich 1988; from Förstner 1995a)

	Reduced water	Oxidized water
Ammonia	125 mg/l	< 3 mg/l
Iron	80 mg/l	< 3 mg/l
Nitrate	< 3 mg/l	120 mg/l
Zinc	< 10 μg/l	5000 μg/l
Cadmium	< 0.5 μg/l	80 μg/l

In municipal solid waste landfills, initial conditions are characterized by the presence of oxygen and pH-values between 7 and 8. During the subsequent "acidic anaerobic phase", the pH drops to a level as low as 5 because of the formation of organic acids in a more and more reducing milieu, and concentrations of organic substances in the leachate are high. In a transition time of 1 to 2 years, the chemistry of landfill changes from acetic to methanogenic conditions. Typically increased concentrations of metals have been found for iron, manganese and zinc in leachates during the acidic decomposition phase compared with the methanogenic phase (Table 10.3).

There is not much experience with landfill evolution subsequent to the initial 30 years. What could happen is that the landfill is again oxidized. It has been inferred that oxidation of sulphidic minerals by intruding rain-water may mobilize trace metals. The impact on the underlying groundwater could be even higher if a chromatographic process, involving continuous dissolution and reprecipitation during passage of oxidized water through the deposit, were to preconcentrate critical elements prior to final release with the leachate (Förstner 1995b).

Table 10.3. Long-term chemical developments in reactor landfills (Förstner 1994)

	Acidogenic phase	Methanogenic phase	Post-Methanogenic phase
BOD_5 (mg/l)	13,000	180	*Organic substance*
COD (mg/l)	22,000	3,000	*to form colloids?*
pH	6.1	8	*Sulfide oxidation?*
Calcium (mg/l)	1.200	60	*Interactions between H^+, Fe^{3+}, Ca^2*
Iron (mg/l)	780	15	*and solid matrices will*
Manganese (mg/l)	25	0.7	*increase mobility of trace*
Zinc (mg/l)	5	0.6	*elements*

In the study on the material management within a fictitious region "Metaland" (Baccini and Brunner 1991), an estimation was made by Baccini et al. (1992) on the release of contaminants from landfills subsequent to a 50-year reactive period, for the case that all barriers will have failed at this time. The population in Metaland was assumed as being 1 million on an area of 2500 km^2, each producing 40 tonnes of waste within a 40-year time-span. Table 10.4 indicates, that a very significant input of TOC (and of ammonia) with an annual increase of 50% compared to the uncontaminated groundwater will occur. It has been suggested that this input will last over a period of approximately 1000 years.

Table 10.4. Effects of landfill leachate on chemical composition of a groundwater reservoir in the Metaland model study (Baccini et al. 1992)

	C_{org} [mg/l]	Cl [mg/l]	Zn [μg/l]	Cd [μg/l]	Hg [μg/l]
Mean leachate concentration after 50 years land operations	600	500	600	2	0.1
Mean concentration in uncontaminated groundwater	0.5	3	5	0.02	0.005
Mean annual increase of concentration in groundwater	0.24	0.2	0.24	0.0008	0.00016
Mean annual increase in %	50%	7%	5%	4%	3%

Organic substances as main components of so-called reactor landfills may affect subsequent systems in several ways as:

- A driving force and important regulating factor for the cycles of major, minor and trace elements; biochemical processes are playing a major role
- A source material for the formation of colloids, a "third phase" by which an intensive transport of less soluble substances is possible
- A mediator for dissolution processes and thereby enhancing mobility of trace elements, mainly by complexation processes

Processes involving degradation of organic matter typically exhibit non-linear and delayed developments. Different from mineralogical systems, these effects cannot sufficiently be described, modelled and predicted. This is even worse, as important decisions on waste technologies would need information about the long-term behaviour of critical pollutants.

10.4 Geochemical Concepts for Metal Pollutants in Solid Wastes

Four geochemical concepts demonstrate the advantage of a long-term strategy compared to the traditional approaches in waste management: (i) the mobility concept, (ii) the concept of capacity-controlling properties, (iii) the concept of final storage quality and (iv) the concept of coupled geochemical cycles, the first being relevant for process studies, the second for effect evaluations, the third for problem solutions and the forth as an integration of geochemical approaches to processes, prognoses and problem solutions.

10.4.1 Mobility Concept

Among the criteria to assess which element or elemental species (beside its toxic potential) may be of major concern in ecological evaluations, one question deserved primary attention (Andreae et al. 1984): "Is the element mobile in geochemical processes, mainly because of its volatility or solubility in water, so that the effect of geochemical perturbations can propagate through the environment?"

Typically for systems involving solution–solid interactions, *mobility* is given by accelerating and retarding factors and processes. The former influences comprise effects of pH-lowering, redox changes, inorganic and organic complexation and microbially mediated species transformations such as biomethylation. Among the spectrum of "barriers", physical processes include adsorption, sedimentation and filtration; chemical barriers comprise mechanisms such as complexation and precipitation; biological barriers are often associated with membrane processes, which can limit translocation of metals, for example, from plant roots to the shoots and fruits.

10.4.2 Final Storage Quality

The final storage approach is one way to develop and control landfills on a conceptual basis. It has been defined by the Swiss Federal Government in 1986 (Anonymous 1986b) and received wider attention by the book edited by Peter Baccini on Landfills – Reactor and Final Storage (Baccini 1989): "Landfills with solids of final storage quality need no further treatment of emissions into air and water".

Solid residues with final storage quality should have properties very similar to the earth's crust (natural sediments, rocks, ores, soil; see Table 10.5). This can be achieved in several ways, e.g. by assortment or thermal, chemical and biological treatment.

Table 10.5. Comparison of inventories of chemical components in the two landfill alternatives and in the earth's crust (Förstner et al. 1989)

Reactor landfill	Final storage	Earth's crust
Major solid constituent		
Solid "inert" waste	Silicates, oxides	Quarz, Fe-oxide clay, carbonates
Putrefactive waste	[Gypsum, NaCl][a]	(Gypsum, NaCl)
Grease trap waste	(Char)[b]	Kerogenic compounds
Minor solid constituents		
Organic micropollutants	Organic micropollutants	—
Metals in reactive chemical forms	Metal-bearing minerals, mainly oxides	Metals mainly in inert forms
Dissolved constituents		
Protons, electrons	(Protons)	(pH: acid rain)
Organic compounds	(Organic residues)	(Humic acids)
Dissolved salts	[Dissolved salts][a]	(Dissolved salts)

[a] Partial extraction during pretreatment
[b] Minor constituent

10.4.3 Storage Capacity-Controlling Properties

This conceptual approach has been developed in the framework of the concept of "chemical time bombs" (a waste deposit or contaminated hot-spot which initially appears to be relatively harmless, but which can eventually have disastrous environmental effects as toxic contaminants are released). The effects of these time bombs are non-linear and delayed, e.g. toxic metals can "break through" once the specific buffering capacity of a sediment or soil system has been surpassed. To make the scientific objectives clearer, it is useful to distinguish between two different mechanisms (Stigliani 1992): the first is direct saturation, by which the capacity of a soil or sediment for toxic chemicals becomes exhausted. The second way to "trigger" a time bomb is through a fundamental change in a chemical property of the substrate that reduces its capacity to adsorb (or keep adsorbed) toxic materials.

10.4.4 Concept of Coupled Geochemical Cycles

The conceptual approach of "storage capacity-controlling properties" forms the link between the geochemical cycles comprising driving forces such as organic matter/redox-sensitive elements and cycles of mobilizable micropollutants. This concept of coupled cycles has been designed by Salomons (1993; Fig. 10.1).

Biogeochemical cycles involving redox transformations usually are relatively slow compared to physico-chemical processes such as dissolution or desorption. Speed of transmission of mobilizing agents like dissolved organic carbon, salt ions or protons may be reduced by capacity-controlling properties such as cation exchange and pH buffer capacity; in this sense "R" as the difference between the outer and inner transmission wheel reflects retardation mechanisms, mainly of the solid matter, of the system. In the present version, the system includes an additional cycle of accumulation organic matter (see Fig. 10.1).

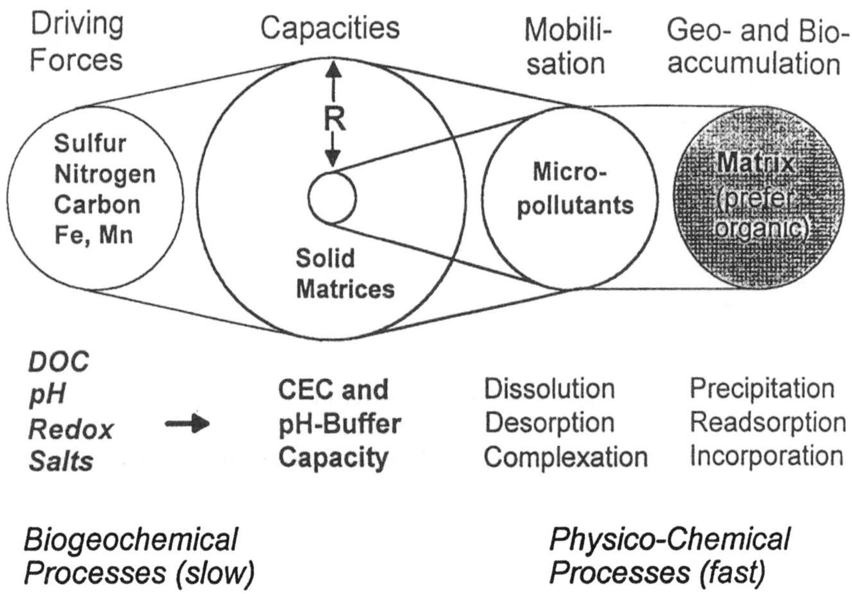

Fig. 10.1. Extension of Salomons' (1993) concept of coupled geochemical cycles

Such specific mechanisms have been demonstrated by experiments in a multi-chamber device, where competition between various target substrates can be studied (Calmano et al. 1988). In fact, the dominant role of organic matrices in the binding of metals like cadmium or copper and of organic micropollutants is of particular relevance for the transfer of these compounds into biological systems. It can be expected that even at relatively small percentages of organic substrates, these materials are primarily involved in metabolic processes and thus, may constitute the major carriers by which micropollutants are transferred within the food-chain.

A typical example of coupled cycles leading to non-linear, delayed mobilization of pollutants is the so-called "split of sulphate" in sedimentary redox cycles. Fig. 10.2 (after van Breemen 1987) indicates that acidification in alternating aerobic and anaerobic systems proceeds in two stages: the first stage is characterized by the reduction of sulphates, for example in tidal flats. At the same time, organic matter is degraded. Most of the sulphide formed is fixed in the sediment as FeS or FeS_2 . Whereas, the acid-producing potential is fixed in the sediment, the acid-neutralizing capacity, in the form of hydrogen carbonate, is mobile and can be flushed away. After the next aeration and oxidation, extreme acidification of the system can take place (Förstner 1995a).

Aerobic conditions <——> Anaerobic conditions

4 $MeSO_4$	+	Fe_2O_3	"CH_2O" ——>	2 FeS_2	+	4 $Me(HCO_3)_2$
4 H_2SO_4	+	Fe_2O_3	O_2 <——	Acid Producing Potential (APP) *solid*		Acid-Neutralisation Potential *dissolved* ——>

Fig. 10.2. "Split" of sulphate in the redox cycle (after Van Breemen 1987)

In Fig. 10.3 possible implications on metal concentrations in sediments have schematically been designed. After deposition and during storage in reduced coastal sediments, absorbed metals are transferred to sulphide-bonding forms. During erosion processes, oxidation can take place involving remobilization of a certain proportion of the sediment-associated metals, which may subsequently be readsorbed in part to particulate matter. It can be expected that through repeated but irregular cyclic processes of erosion and deposition, a significant release of metals from sediments will take place.

From enclosure experiments in Narragansett Bay, Hunt and Smith (1983) estimated that by such mechanisms the anthropogenic proportion of cadmium in marine sediments is released to the water within approximately 3 years. For remobilization of copper and lead, approximately 40 and 400 years, respectively, is needed, according to these extrapolations.

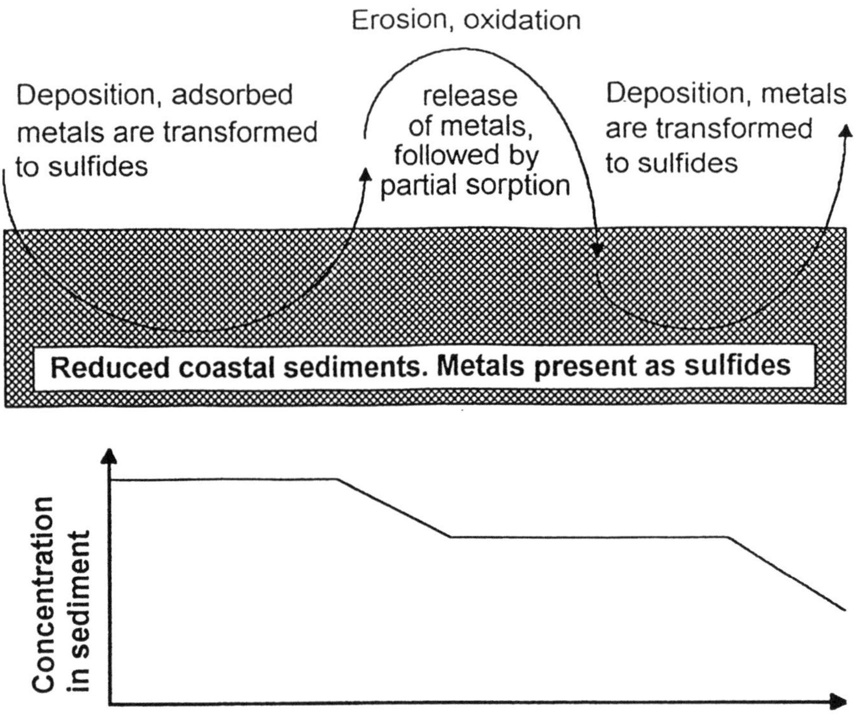

Fig. 10.3. Decrease in metal concentrations in coastal sediments through cyclic processes of erosion and deposition (Salomons 1995)

10.5 Long-term Prognosis of Pollutants Mobility

A general experience is that, the further into the future the process of pollutant release is, the less reliable a prediction of its effects, for example on groundwater quality, will be. Regarding the situation of leachates originating from different sources, a more complex picture is presented (Fig. 10.4).

There are relatively fast and intensive signals from dredged material sedimentation polders (see Table 10.2) and from spoil heaps formed by residues from ore and coal mining. Typical evolution of leachates from traditional municipal solid waste disposal has been presented in Table 10.3. Pretreatment with biological–mechanical procedures could remove reactive organic compounds (see Table 10.4), but will not principally eliminate the problem of long-term production of leachates, which have to be treated. Reliable long-term perspectives are only provided by widely inert residues, where organic matter has been destructed by thermal treatment (see Table 10.5, p. 161). Reliability of long-term predictions will be improved when in-depth knowledge is available on primary

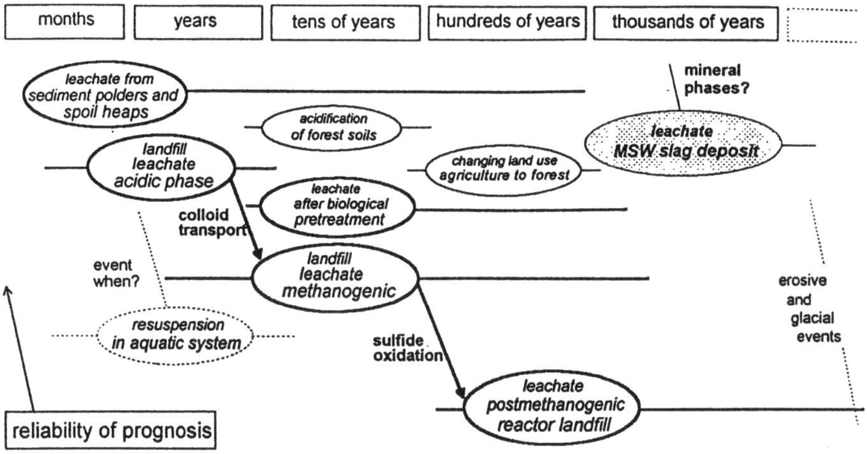

Fig. 10.4. Temporal sequence of different types of leachates with an estimation of prognostic reliability regarding effects on groundwater quality, particularly from heavy metals

and secondary mineral phases.

Analytical and experimental procedures as well as technologies related to final storage can be deduced from the concept of interlinkage between major biogeochemical cycles and capacity-controlling properties, as has previously been shown in Fig. 10.1 (Fig. 10.5):

- Transformation of reactive organic substances will result in variations of redox potential, pH-values, as well as concentrations of inorganic ions and dissolved organic carbon; the latter two components may enhance complexation of metals. Regarding the subsequent consideration, acid-producing potentials derived from oxidation of sulphides and organic matter are of prime importance.
- Second are the properties of solid phases, which can be subdivided in properties of the structure and the surface of minerals and solid organic matter, and the conditions in the pore space between the particles. In practice, high priority is given to redox and pH buffer as well cation exchange capacity.
- A third group of information refers to the physico-chemical behaviour of the potential pollutants. Most common are determinations of elution and desorption rates. Much less is usually known about the diffusion properties.

In practice geochemical investigations are mostly aimed for long-term prediction of typical borderline conditions, such as redox, complexation and acid-producing potential. Here, pH-conditions form the master factor for metal-related problems. Therefore, key information on the long-term behaviour of metals can be provided from combined data of specially-designed elution experiments and extrapolations of the acid-producing potential under worst case conditions. Regard-

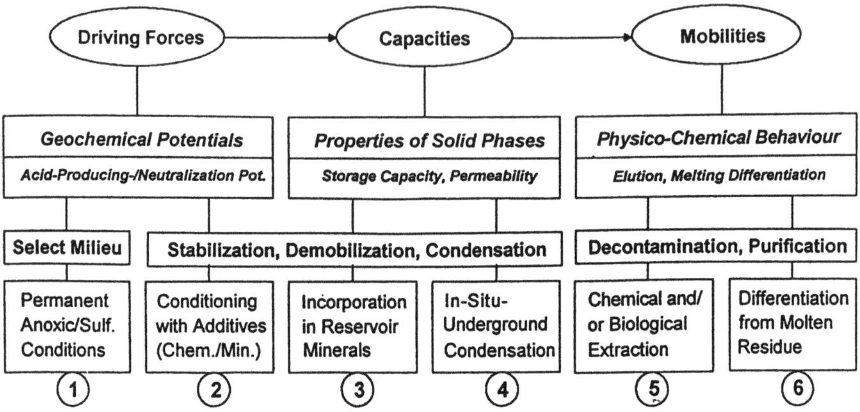

Fig. 10.5. Analytical/experimental methods and treatment technologies derived from the concept of interlinkage of geochemical cycles (see Fig. 10.1)

ing long-term behaviour of organic-rich waste materials, such as reactor landfills, redox buffer systems should be considered as well (Heron and Christensen 1995).

10.5.1 Short- and Long-term Prognosis on Contaminated Sediments

Discussion of available test procedures for short- and long-term prognosis has exemplarily been presented by Förstner (1995a) for sediments: (1) Composition of pore water is a highly sensitive indicator for reactions between chemicals on solid substrates and the aqueous phase which contacts them. While the direct recovery and analysis of water-borne constituents can be seen as a major advantage of this approach, there are several difficulties, particularly arising from the sampling and sample preparation. (2) The advantage of short-term elutriate tests is that especially important parameters can be directly observed and particularly unfavourable conditions can be simulated. Interpretation may become difficult for systems undergoing redox variations, which is usually the case, both in aquatic sediments and land-disposed dredged materials. (3) The undisputed advantage of sequential extraction procedures lies in the fact, that rearrangement of specific solid phases can be evaluated prior to the actual remobilization of certain proportions of an element into the dissolved phase; when proceeding further in the extraction sequences, usually a reduction in prognostic reliability takes place. (4) Simultaneous application of standard sequential leaching techniques can be used for geochemical characterization of anoxic, sulphide-bearing sediments in relation to the potential mobility of critical trace metals (Kersten and Förstner 1991). For determining the acid-producing capacity in anoxic sediments, both the FeS pool (actual APC) and the maximum ferrous sulphide producing capacity upon disposal, reflected by the reactive (reducible) iron concentration, has to be considered. (5) With the pH-stat-technique (Obermann and Cremer 1992), a combined determination of metal releases and buffer capacity can be made at different time intervals. The sum curve of acid consumption provides information on the potential changes of the matrix composition during acidification.

10.5.2 Long-term Metal Release from Industrial Waste Material

In a series of experiments, a number of industrial waste materials of different types, intended for co-disposal in borrow pits, has been studied in a circulation apparatus by the controlled intensification of significant release parameters such as pH-value, redox-potential and temperature (Schoer and Förstner 1987). Figure 10.6 presents the cumulative mobilization of zinc from two different waste compositions, which were studied in experiments at pH 5 for 10 weeks. The addition of the high-pH component is reducing the release of Zn in the initial stage of the experiment. From the sixth week on, however, the release of zinc from this material is distinctly increasing and is finally surpassing the rates from the untreated industrial waste. From these data, it could be argued that the long-term effect of the initial high-pH additive on the stability of trace metals in the disposal mixture is less advantageous than previously assumed. This is due to the oxidation of small concentrations of sulphur-containing compounds. It is also clear from this experiment that, with a standard leaching test, over one hour or one day, such effects would not have been detected.

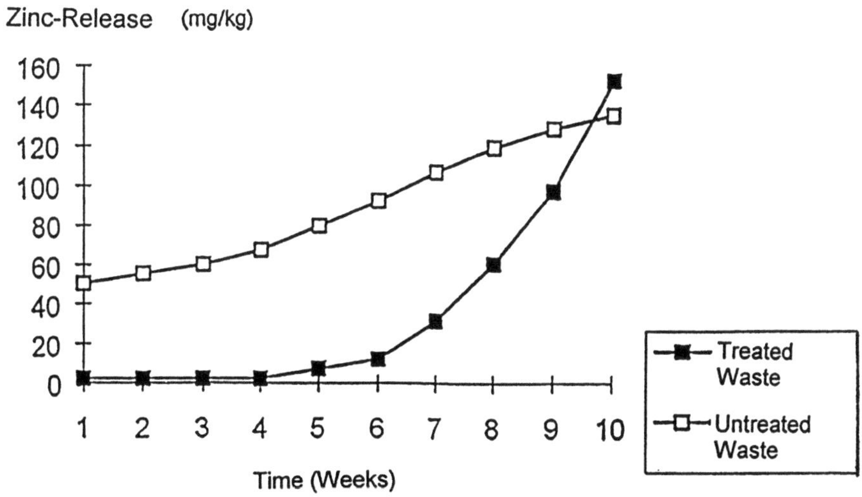

Fig. 10.6. Cumulative mobilization of zinc from original waste, *open squares*, and subsequent to stabilization by high-pH additive, *solid squares* (Förstner et al. 1991)

10.5.3 Long-term Chemical Behaviour of Incineration Products

Incineration of 1 tonne of municipal solid waste provides approx. 250 to 350 kg of bottom ash and slag, 20 to 40 kg of precipitator dust, and – depending from the purification procedure – 8 to 45 kg of reaction products, among others gypsum and related components. Bottom ash and slag until now is not treated to the same extent as precipitator dusts, however, there is a tendency to introduce directives for handling these materials as well, at least in respect for their reutilization in road construction etc.

Studies of Baccini and colleagues have shown, that under long-term considerations even products like incineration bottom ash slag are not really inert. After the release of chloride and sulphate, two phases can be distinguished (Fig. 10.7): part of the calcium content, which was formerly present as hydroxide, is now carbonatized. Iron and aluminium are transformed to oxides. On the long run, the carbonate buffer in the slag will be emptied. This will be correlated by relative low pH-values. Compared to the former conditions, elevated trace metals concentrations in the leachate can be expected.

The long-term release of calcium, chloride, sulphate and heavy metals from bottom ash and slag has been estimated by the Swiss EKESA-Project (1992) and

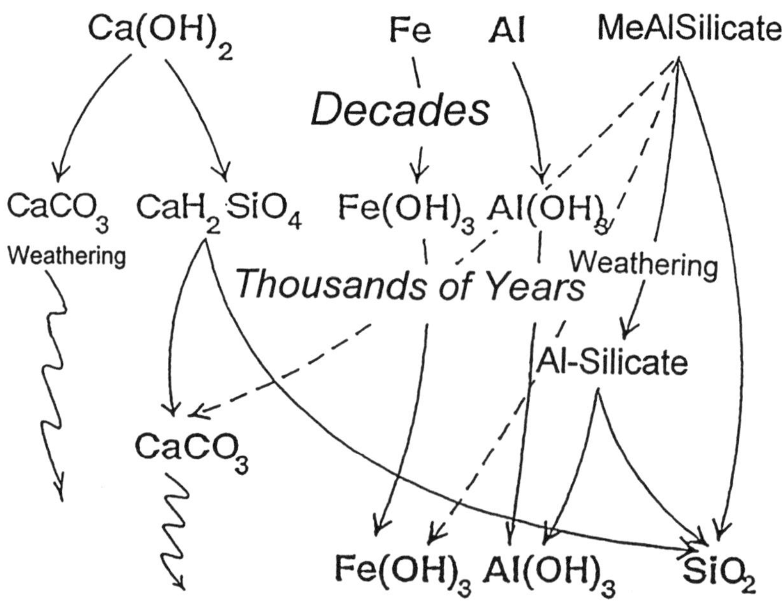

Fig. 10.7. Long-term chemical evolution of municipal solid waste incineration slags on mono-landfills (Baccini et al. 1993)

Kersten et al. (1995). *Phase 1* in Table 10.6 is characterized by very high concentrations of easily soluble components such as chloride and sulphate; the slag is not yet carbonatized; portlandite and gypsum may influence leachate composition to some extent. *Phase 2* follows after emptying the contents of chloride and sulphate in the slag. Part of the calcium content, which was formerly present as hydroxide, is now carbonatized. It can be expected that cadmium and lead are present as carbonates, since these compounds are less soluble than the respective hydroxides. Zinc concentrations are widely unchanged; chromium contents, however, which cannot find a suited host crystal lattice in this phase, may strongly increase in the leachate of phase 2 compared to the conditions in phase 1 (Kersten et al. 1995). *Phase 3* exhibits a far-reaching emptying of the carbonate buffer in the slag. This is correlated by relative low pH-values, which is just buffered by the still-present oxides and silicates. Compared to phase 2 elevated trace metals concentrations can be expected.

Table 10.6. Development of leachate concentrations of selected elements in a slag deposit (EKESA-Project 1992; Kersten et al. 1995)

	Phase 1	Phase 2	Phase 3
pH	8.3 and higher	7.3–8.3	< 5–6
Calcium	520 mg/l	16–60 mg/l	< 16 mg/l
Chloride	100–5000 mg/l	< 100 mg/l	< 100 mg/l
Sulfate	100–5000 mg/l	< 100 mg/l	< 100 mg/l
Zinc	4 μg/l	unchanged	high
Chromium	13 μg/l	Phase 1 times 10^6	as Phase 2
Duration	tens of years	hundreds of years	> 3000 years

Long-term prediction on chemical variations in quasi-inert waste deposits has been demonstrated by Belevi and colleagues (1992) for slags from municipal solid waste incineration (Fig. 10.8). The acid-neutralizing capacity has been measured to be approximately 2 moles per kilogram of solid matter. The slag residues still contain approximately 2% of organic carbon. If we assume, that half of it is still reactive to produce CO_2, this would have used up 1 mole of buffer capacity. One mole would be left and the situation can be considered as being safe for long time. Effects from acid rain can be calculated as follows (Fig. 10.8): assuming the input of acid precipitation of pH 4 and 1000 mm per year, the interaction with a slag layer of a 10-metre thickness for 1000 years would only produce 10 millimole protons per kg of slag; this is 100 times less than the potential internal production of acidity.

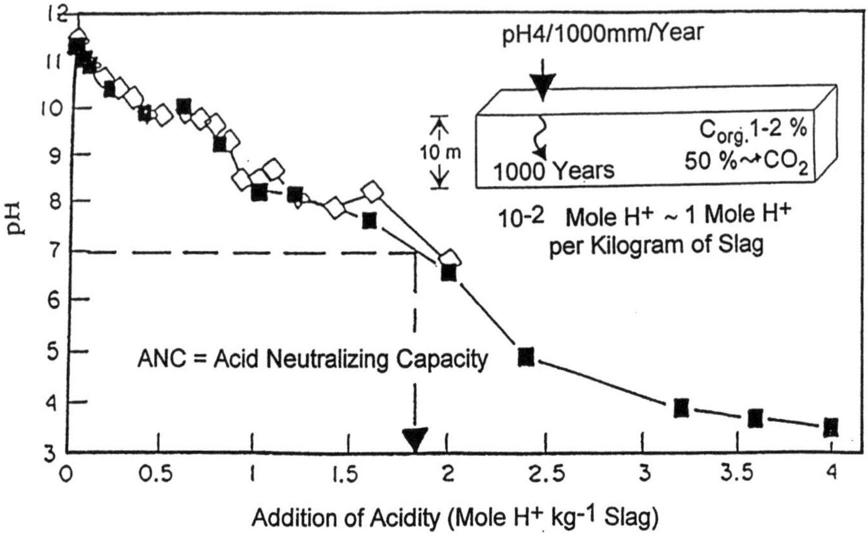

Fig. 10.8. Titration of bottom ash samples. Full signature: Extraction period 48 h. Open signature: 197 h (after Belevi et al. 1992)

10.6 Geochemical Engineering: Technological Problem Solutions

The term "Geochemical Engineering" has been proposed first by Wim Salomons – in the foreword to a book on the treatment of dredged material and mine wastes (Salomons and Förstner 1988). The term was used in the meaning, that geochemical principles are applied in certain technological problem solutions, such as stabilization and solidification of contaminated matrices. Two years later, Schuiling (1990) has described the new field in more detail. His considerations were in the direction of detoxification and decontamination, that is, the extraction of pollutants from a solid matrix by means of degradation or elution.

The two lines of geochemical engineering – stabilization and purification – have been treated in a more recent article by Voronkevich (1994) on engineering geochemistry. There is a strong connection to the earlier work of Soviet geochemists such as Vernadsky and Fersman and their co-workers in the thirties, who have introduced the expression "geochemistry of technogenesis" for defining the geochemical consequences of industrial and agricultural activities.

The two geochemical fields of application can directly be connected with the major fields of environmental technology – civil engineering on the one side and chemical engineering on the other side. On the one hand, we have techniques such as solidification, stabilization, enclosure walls to inhibit escape of pollutants from a deposit, encapsulation (more in a micro-scale), vitrification and brickmaking, for example. Chemical engineering techniques are separation by mechanical, chemical, thermal, magnetic and electrolytic procedures, and there are various methods for reenrichment on surfaces, as precipitation, sorption, etc.

Starting from the three groups of prognostic methods given in Fig. 10.5 – acid-producing potential, storage capacity and elution – various techniques can be derived for treatment, safeguarding and sanitation of solid wastes and contaminated soil. Regarding the two extreme possibilities, the waste material can either be stored under environmental conditions, where no release of pollutants can occur, or the treatment is done in a way, that critical components are extracted from the residual matrix. In the latter case, reuse of the matrix can be considered; however, there might be stricter rules than for disposal of the residue.

The discussion actually is focussing on the range between the two extremes. For the most favourable case, stabilizing additives affect permanent fixation of pollutants under realistic conditions which equals "immobilization". If dispersion of pollutants is significantly reduced, then the term "demobilization" would be appropriate. During stabilization or demobilization, different approaches can be taken, which also can be combined:

- Sufficient long-term buffer capacity can be provided, according to the environmental conditions.
- New formations of "reservoir minerals" can incorporate potential pollutants in their internal structure.
- Permeability for dissolved contaminants can be reduced by secondary mineral precipitations or by soft gels injected into the pore space of the waste body.

In the following, six groups of problem solutions will be presented. In view of the large number of examples, different cases could have been selected.

10.6.1 Storage Under Permanent Anoxic Conditions

Regarding the various containment strategies, it has been argued that upland containment (e.g. on heap-like deposits) could provide a more controlled management than containment in the marine environment. However, contaminants released either gradually from an imperfect impermeable barrier or catastrophically from failure of the barrier could produce substantial damage (Kester et al. 1983). On the other hand, near-shore marine containment (e.g. in capped mound deposits) offers several advantages, mainly with respect to the protection of groundwater resources since the underlying water is saline and inherent chemical processes are favourable for the immobilization or degradation of priority pollutants.

In a review of various marine disposal options, Kester et al.(1983) suggested that the best strategy for disposing of contaminated sediments is to isolate them in a permanently reducing environment. Disposal in capped mound deposits above the prevailing sea-floor, disposal in subaqueous depressions and capping deposits in depressions provide procedures for contaminated sediment (Bokuniewicz 1982). In some instances it may be worthwhile to excavate a depression for the disposal site of contaminated sediment which can be capped with clean sediment. This type of waste deposition under stable anoxic conditions, where large masses of polluted materials are covered with inert sediment became known as "subsediment-deposit". The first example was planned for highly contaminated sludges from Stamford Harbour in the Central Long Island Sound following intense discussions in the US Congress. The last proposal as yet is that of Müller (1995) with disposal in deep anoxic structures like the Black Sea.

The Black Sea consists of a surface layer of 150 to 250 metres of oxic water (Fig. 10.9). Below the halocline, the water body, with a maximum depth of 2200 m, is anoxic and containing methane and hydrogen sulphide. Both these components also occur in the sediment. Metal sulphide minerals were formed in the sediment, exhibiting low solubilities, and this would prevent release of critical trace elements. The hydrographic conditions can be considered as stable for long period of time, and the worst change would be a drastic depression of the world oceans. But even then, anoxic conditions would be preserved for geologic times due to the content of organic substance.

In fact, even if this possibility will not be used due to political reasons, the anoxic geochemistry constitutes constitutes favourable conditions for the immobilization of toxic metals in large-volume waste materials. This is valid both with respect to environmental safety and economic considerations. Metal sulphides in particular have very low solubilities compared with those of the respective carbonate, phosphate and oxide compounds. For the extreme case of copper, there

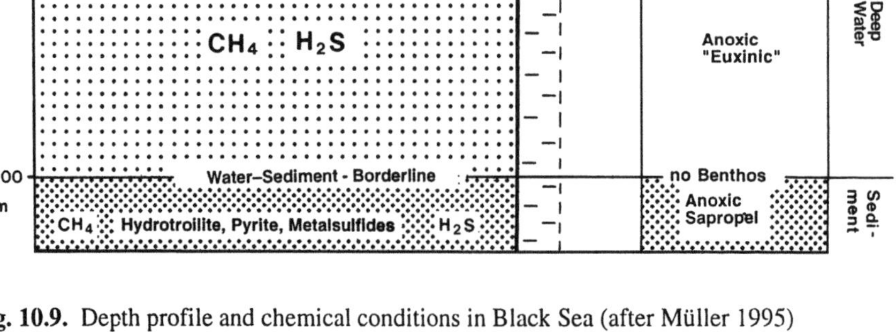

Fig. 10.9. Depth profile and chemical conditions in Black Sea (after Müller 1995)

are 12 orders of magnitude between the solubility of sulphide and hydroxide, and even for the case of zinc, there are still 7 orders of magnitude (Ehrenfeld and Bass 1983). Sulphide bonding will remain a favourable alternative to other forms of metal fixation.

In principle, the chemical stability of harbour sludge deposits, such as at Rotterdam and Hamburg (Fig. 10.10), is based on the low solubility of metal sulphides as well. The major difference is less mechanical stability of these structures compared with the deep anoxic marine basins, such as the Black Sea (many fjords provide similar conditions). Compaction and erosion are the problems at the man-made disposal sites.

Figure 10.10 shows cross-sections of the dredged material deposits of Rotterdam and Hamburg. Rotterdam needs space for 20 million cubic metre per year, about ten-fold the mass of the Hamburg sludge. It soon became obvious, that no disposal site can be found on land for this mass of sludge. Therefore, an artificial peninsula was constructed in front of the mouth of Rhine and Scheldt rivers. A 20-metre deep pit was excavated on the sea-floor, and a dike was formed with the sand material. The effective depth of the basin is 32 m, and the water-rich sludge, which is dredged from the harbour is just spilled from the pipeline head into the basin. Due to consolidation of the material, approximately 150 million

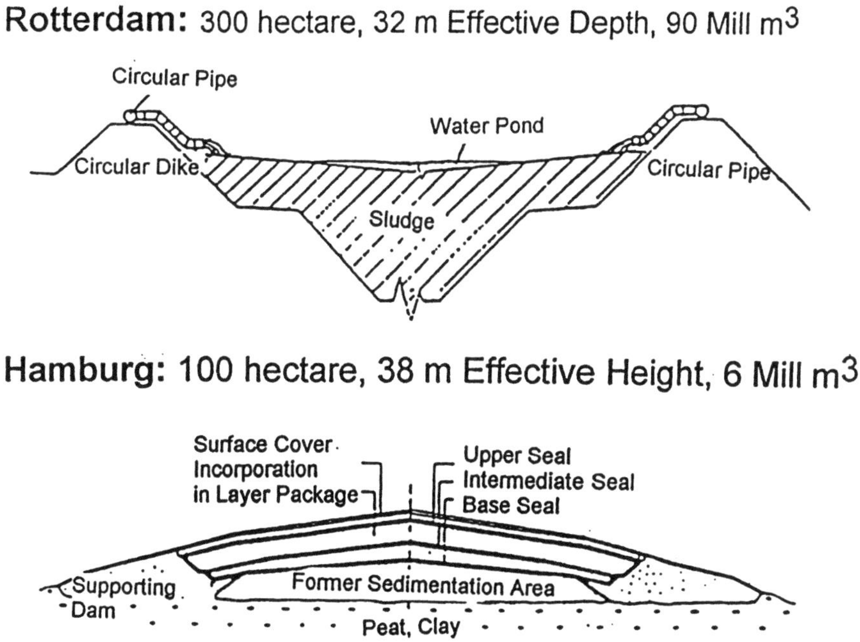

Fig. 10.10. Cross-sections of dredged-material disposal sites of Rotterdam (off Rhine/Meuse) and Hamburg (Francop site) (after Donner et al. 1994)

cubic metres will eventually be stored in the peninsula, instead of the originally planned 90 million cubic metre of wet sludge.

At the Hamburg harbour site, separation of the fine fraction is performed by cyclonage and up-stream classification before disposal. The deposition is made in alternating layers of 1.5 m fine-grained material and 30 cm of sand, the latter to drain excess water from the deposits. The rest of the separated sand fraction is used for the supporting dam. As we will see later, the costs of the Hamburg procedure are much higher than at Rotterdam, because of the efforts to separate and dewater the material. Capacities of both structures will be exhausted in about 10 to 15 years.

10.6.2 Chemical Stabilization by Additives

In general, solidification/stabilization technology is considered a last approach to the management of hazardous wastes. The aim of these techniques is a stronger fixation of contaminants to reduce the emission rate to the biosphere and to retard exchange processes. Most of the stabilization techniques aimed for the immobilization of metal-containing wastes are based on additions of cement, water glass (alkali silicate), coal fly ash, lime or gypsum (Malone et al. 1982; Goumans et al. 1991).

Laboratory studies on the evaluation and efficiency of stabilization processes were performed by Calmano et al. (1988) on Hamburg harbour mud, using acid titration curves for limestone and cement/fly ash stabilizers. Best results are attained with calcium carbonate, since the pH-conditions are not changed significantly upon addition of $CaCO_3$. Generally, maintenance of a pH of neutrality or slightly beyond favours adsorption or precipitation of soluble metals (Gambrell et al. 1983). On the other hand, it can be expected that both low and high pH-values will have unfavourable effects on the mobility of heavy metals.

In general, microscale methods, e.g. formation of mineral precipitates in the pore space of a waste body, will be employed rather than using large-scale enclosure systems such as clay covers or wall constructions (Wiles et al. 1988). An overview on various fields of environmental research and management to which mineralogical methods can be successfully applied has been given by Bambauer (1991).

10.6.3 Incorporation of Pollutants in Reservoir Minerals

There is a new field of stabilization techniques, which uses the properties of certain minerals to incorporate and store critical components in their crystal lattice. Such techniques are mainly applied on industrial process residues. Pöllmann (1994) has demonstrated two different ways to the formation of a *stabilisate*, where pollutants can be stored for a long period of time (Fig. 10.11). In the first

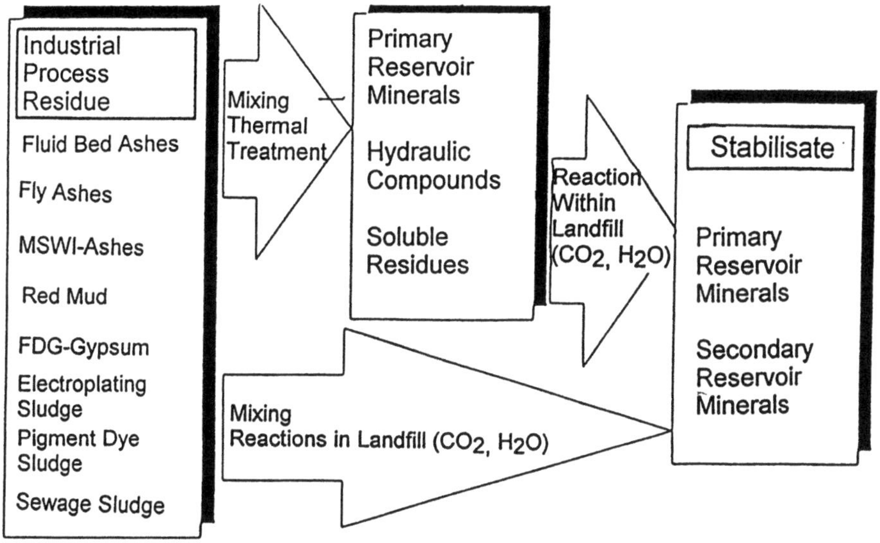

Fig. 10.11. Procedural steps for the formation of reservoir minerals from industrial mineral residues (Pöllmann 1994)

case, primary reservoir minerals, which exhibit no hydraulic reaction, act as widely inert filling substances for hydraulic formations within the landfill. In case number two, a second generation of minerals, which also can incorporate pollutants from the water phase, fills the pore space. By such condensation processes, water permeability is reduced, and consequently also the dispersion of pollutants by convection.

Examples of reservoir minerals are (Pöllmann 1994):

Ellestadite	$\mathbf{Me}_{10}[\mathbf{Cl}_2/(SiO_4)_3(\mathbf{SO_4})_3]$
Sodalite	$\mathbf{Me}_8[Al_{12}O_{24}] \bullet (\mathbf{XO_4})_2$
Ettringite	$[Ca_6Al_2(OH)_{12} \bullet 24\,H_2O]^{6+}[(\mathbf{SO_4})_3 \bullet 2H_2O]^{6-}$
Me/Me-Hydroxisalt	$[\mathbf{Me_6}Al_2(OH)_{16}]^{2+}[CO_3 \bullet nH_2O]^{2-}$
Ca/Al-Hydroxisalt	$[Ca_4Al_2(OH)_{12}]^{2+}[(OH)_2 \bullet nH_2O]^{2-}$

Experimental studies of the processes taking place with mixed residues from lignite coal incineration indicate favourable effects of incorporation of both chloride and heavy metals in newly formed minerals. Ettringite in particular, can act as "storage minerals" for chloride and metal ions. The former may be incorporated at up to 4 kg $CaCl_2$ per m^3 of the mineral mixture. Calcium-silicate-hydrate phases may be formed in a subsequent process, and by filling further pore space these minerals can significantly reduce permeability of the waste body for percolating solutions. Experimental studies of the leachability of salts and trace elements from samples of "*stabilisates*", with a pressure-filtration method, indicate relative high rates of release for sulphate ions, but not for zinc and cadmium in the eluate (Bambauer 1992).

10.6.4 Compaction by Grouting Gels

Injection of silica-rich solutions acts in the same direction as the before-mentioned procedures of chemical stabilization, namely by fixation of metals within a favourable pH-range. Grouting gels, in addition, by filling the pore space, will significantly reduce permeability for dissolved pollutants.

Water glass is produced by co-melting of stoichiometric amounts of sodium carbonate and silica. The newly formed alkali silicates are water-soluble under pressure and exhibit strong basic reaction. In contact with atmospheric CO_2 and decreasing pH polymerization is enhanced. Eventually a three-dimensional network of a so-called soft-gel is formed, which agglutinates the soil or waste particles. Stability of the soft gel structure is preserved at pH less than 7. Full dryness affects formation of solid silica.

Experiments have been made by Belouschek et al. (1995) with different mixtures of mass residual materials such as sewage sludge, recycling sand and fly ash, which were treated with water glass. Laboratory data were compared with field applications. It is shown, that in all studied cases permeability is much lower than prescribed by landfill guidelines. For all mixtures, permeability after 3 months of application is equal or less than in the experiments.

10.6.5 Chemical and/or Biological Extraction (Overview)

Many of the procedures applied in this field are modifications of chemical engineering techniques, mainly in the area of solid–solid or solid–liquid separation. The more advanced techniques are washing procedures, of which the fine-grained residues are eventually deposited. With respect to acid extraction procedures, there are many experimental designs, but few practical applications. Electrochemical procedures are in progress, in particular for in-situ-reclamation of metal-rich areas.

Relatively close to the term "geochemical engineering" is the technique of bacterial leaching. The technique is known for a long time from the treatment of metal-poor ores, where hydro- or pyrometallurgy is no more economically feasible. During these processes – mediated by specialized bacteria – oxidation of sulphur and sulphides produces sulphuric acid, and metals are mobilized such as copper.

Figure 10.12 gives the scheme of a pilot plant, which has been designed by the Environmental Research Centre at Leipzig. With this technique, contaminated river sediments should be treated and decontaminated to an extent, that they can

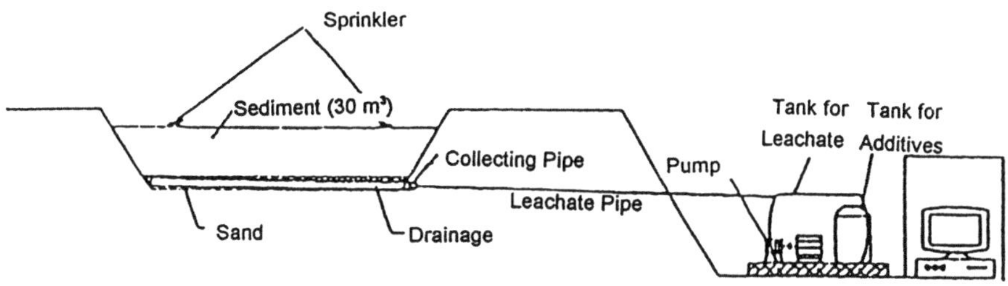

Fig. 10.12. Bacterial leaching pilot plant (Seidel et al. 1995)

be disposed in former open pits from lignite coal mining. These large areas in eastern Germany have to be recultivated during the coming years.

10.6.6 Melting Differentiation of Incineration Products

Recent advances in waste incineration mainly concern the application of high temperature techniques for remelting precipitator ashes and flue gas purification products (Faulstich and Zachäus 1992). The flame chamber technique is in operation already with more than 30 plants, mainly in Japan. Glass melt and electric melting processes are widely developed. More recent development apply gasification at low temperature, which can be combined with a high-temperature step to produce molten residues. Very high temperature is involved in the plasma melt and electric arc technology.

An example is the RedMelt-process, designed by Faulstich and colleagues (1992), which – in a reductive milieu – affects differentiation into three phases: (1) metal-rich (Fe, Cu) bottom phase, (2) "condensate" phase with enrichment of chlorine, zinc, lead and cadmium in the flue gas and (3) relatively metal-poor silicate phase. This system looks quite similar to a blast furnace, where the same type of differentiation takes place into three phases.

All three fractions can be utilized. Starting at the smaller fractions, there is at first the condensate, which is accumulated in the flue gas control devices. Here, metals can be recovered and transferred to a base metal smelter. Chloride can be reused in aluminium industry or – as already practised – for chlor-alkali-electrolysis, whereby chlorine is recycled.

The bottom fraction is enriched in iron, copper and other valuable components. Scrap iron can also be separated before applying the high-temperature step. Various techniques are available for reuse of these residues, such as blast furnace in steel works or shaft-furnace conversion of the molten material.

Finally, the silicate phase, forming 80 to 85% of the total melting product, is characterized by low metal contents. Therefore, it can well applied in road construction as an additive to concrete and as sand substitute. However, it has been inferred that it would better to have been used to replace other products which consume much energy for the melting process – for example mineral fibres, pumice and foam glass. In this case, the molten silicate fraction from the high-temperature process should directly be transferred into such products, without cooling and remelting.

10.6.7 Comparison of Costs of Selected Procedures

Of the parameters listed in Table 10.7, costs form the master factor, with subfactors being "residue" (in kg/tonne), "energy" (kWh/tonne) and "consumption of chemical" (kg/tonne).

Disposal of dredged material requires from less than ~$3.50 US for direct pipeline transfer (2 km) into an artificial coastal basin (Rötterdam) to about $100 US after preparation, i.e. classification and dewatering (Hamburg). Using relatively simple classification/washing procedures about $100 to $300 US will be needed, depending from disposal costs of the more or less critical residue. Regarding decontamination, costs for acid extraction have been estimated from $170 to $380 US. While the costs can be subdivided approximately half and half for application of chemicals and for disposal of the residues, it seems that the use of complexing agents such as nitrilotriacetic acid in these processes may have some advantages. In-situ electrochemical treatment only produces small masses of residues. However, high costs for energy and the situation of the aquifer may render the process non-competitive compared to other techniques. Finally, the thermal melting method exhibits major advantages, in that, the product can be reutilized as demonstrated before. Due to the high energy consumption, such processes

Table 10.7. Comparison of costs for selected procedures of solid waste treatment (modified after Roos 1995)

	Mono landfill	Washing (water)	Acid extraction	NTA-extraction	Electro-chemical	Thermal (melting)
Residue (kg/tonne)		300–500	120–280	120–320	small	20–40
Energy (kWh/tonne)		10–20	10–20	10–20	200 (el.)	> 1,000
Chemicals (kg /tonne dry material)			500 HCI 150 CaO	70 HCI 80 CaO 9 NTA	100 HCI 90 CaO	
Costs (US$/t) [a]	*<3.5* [b]*-90* [c]	*100–300*	*240–350*	*240–310*	*280–380*	*~ 650*
Application	Dredged material	Soil, Sludges	Residues, Sludges	Residues, Sludges	Soil, Sludges	Fly Ash, Residues

[a] on the basis of 1 Deutschmark = 0.67 US$
[b] Rotterdam (directly via pipeline)
[c] Hamburg (after mechanical treatment)

may only be applied for a small and extremely critical proportion of waste residues, namely precipitator ashes from solid waste incineration processes.

Purification of contaminated solids generally increases their potential for reutilization and should, therefore, have priority compared to any type of conditioning or disposal. So far this is the theory. In practice, after a closer look, we will mostly detect, that purification or decontamination is connected with such high consumption of resources, that there is no ecologic and economic sense for applying these procedures on a larger scale.

10.7 Outlook

In the technological sequence from "natural resource" to "final product" via raw and base material, as well as initial and interim product, there is a general increase in the resource's value. However, simultaneously consumption of material and energy resources, emissions to water and air, as well as production of solid and liquid waste is enhanced. This implies, that recycling should always take place at the highest possible level of complexity to avoid a maximum of emissions and resource consumption. The sum of the argument suggests a long-run imperative favouring an "industrial metabolism" that results in reduced extraction of virgin materials, reduced loss of waste materials, and increased recycling of useful ones (Herman et al. 1989). With regard to remediation techniques, it may well be, that economic considerations will strengthen the popularity of geochemically engineered solutions in large-scale contaminations (Salomons 1995) as well as for large waste masses such as mine residues, dredged materials and filter ashes.

References

Andreae MO et al. (1984) Changing biogeochemical cycles – group report. In: Nriagu JO (ed) Changing metal cycles and human health. Dahlem- Workshop. Springer, Berlin Heidelberg New York, pp 359–373

Anonymous (1980) Groundwater strategies. Environ Sci Technol 14:1030–1035

Anonymous (1986a) National priorities list fact book. HW 7.3, 94 pp. US Environmental Protection Agency, Washington DC

Anonymous (1986b): Leitbild für die schweizerische Abfallwirtschaft (Guidelines for the waste management in Switzerland), Schriftenreihe Umweltschutz No 51. Eidgenössische Kommission für Abfallwirtschaft, Bundesamt für Umweltschutz Bern

Arneth, J-D, Milde G, Kerndorff H, Schleyer, R (1989) Waste deposit influences on ground water quality as a tool for waste type and site selection for final storage quality. In: The landfill – reactor and final storage, Baccini P (ed). Lecture Notes in Earth Sciences, Springer, Berlin Heidelberg New York, 20:399–415

Baccini P (ed)(1989) The landfill – reactor and final storage. Lecture Notes in Earth Sciences 2O. Springer, Berlin Heidelberg New York

Baccini P, Brunner PH (1991) Metabolism of the anthroposphere. Springer, Berlin Heidelberg New York
Baccini P, Bader H-P, Belevi H, Ferrari S, Gamper B, Johnson A, Kersten M, Lichtensteiger T, Zeltner C (1993) Deponierung fester Rückstände aus der Abfallwirtschaft – Endlagerqualität am Beispiel Müllschlacke. Workshop Kartause Ittingen, Sept. 14–16 1993. vdf Hochschulverlag Zürich
Baccini P, Belevi H, Lichtensteiger T (1992) Die Deponie in einer ökologisch orientierten Volkswirtschaft. Gaia 1: 34–49
Bambauer HU (1991) The application of mineralogy to environmental management – an overview. ICAM'91 – Intern Congr on Applied Mineralogy, Pretoria. C.133. vol 1, paper 3
Bambauer HU (1992) Mineralogische Schadstoffimmobilisierung in Deponaten – Beispiel: Rückstände aus Braunkohlenkraftwerken. BWK Umwelt-Special March 1992:29–34
Belevi H, Stämpfli DM, Baccini P (1992) Chemical behaviour of municipal solid waste incinerator bottom ash in monofills. Waste Management Research 10:153–167
Belouschek, P, Kügler JU, van der Sloot HA, Beukems H (1995) Labor- und Praxisuntersuchungen zum Emissionsverhalten von wasserglasvergüteten Dichtsystem im Altlastenbereich und in der Deponietechnik. Abstr V 16 Annual Meeting Water Chemistry Group, German Chemical Society, Hitzacker
Blakey NC (1984) Behavior of arsenical wastes co-disposed with domestic solid wastes. J Water Pollut Control Fed 56:69–75
Bokuniewicz HJ (1982) Submarine borrow pits as containments for dredged sediments. In: Kester DR, Ketchum BH, Duedall IW, Parks PK (eds) Dredged material disposal in the ocean, pp 215–227. John Wiley & Sons, New York
Breemen N van (1987) Effects of redox processes on soil acidity. Neth J Agric Sci 35:271–279
Calmano W (1988) Stabilization of dredged mud. In: Salomons W, Förstner U (eds) Environmental management of solid waste: dredged materials and mine tailings. Springer, Berlin Heidelberg New York, pp 80–98
Calmano W, Ahlf W, Förstner U (1988) Study of metal sorption/desorption processes on competing sediment components with a multi-chamber device. Environ Geol Water Sci 11:77–84
Donner H et al. (1994): Niedersächsisches Elbschlick-Forum. Abschlußbericht 1994. Lüneburg, 185 pp
Dreesen DR, Gladney ES, Owens JW, Perkins BL, Wienke CL, Wangen LE (1977) Comparison of levels of trace elements extracted from fly ash and levels found in effluent waters from a coal-fired power plant. Environ Sci Technol 11:1017–1019
Ehrenfeld J, Bass J. (1983) Handbook for evaluating remedial action technology plans. Municipal Environ Res Lab Cincinnati. EPA-600/2-83-076. Aug 1983
Ehrig H-J (1989). Water and element balances of landfills. In: P Baccini (ed) The landfill – reactor and final storage. Lecture Notes in Earth Sciences. Springer, Berlin Heidelberg New York, 20:83–115
EKESA-Project (1992) Emissionsabschätzung für Kehrichtschlacke. MBT Umwelttechnik Zürich/EAWAG, Abt. Abfallwirtschaft und Stoffhaushalt Dübendorf, 156 pp
Faulstich M, Zachäus, D (1992) Verfahren zur Behandlung von Rückständen aus der Müllverbrennung. In: Faulstich M (ed) Rückstände aus der Müllverbrennung. EF Verlag für Energie und Umwelttechnik, Berlin, pp 1–159
Faulstich M, Freudenberg A, Köcher P, Kley G (1992) RedMelt-Verfahren zur Wertstoffgewinnung aus Rückständen der Abfallverbrennung. In: Faulstich M (ed) Rückstände aus der Müllverbrennung. EF Verlag für Energie und Umwelttechnik, Berlin, pp 703–727

Förstner U (1993) Dispersion of contaminants from landfill operations. In: Petruzzelli D, Helfferich FG (ed) Migration and fate of pollutants in soils and subsoils. NATO ASI Series G: Ecological Sciences. Springer, Berlin Heidelberg New York, 32:435–453
Förstner U (1994) Geochemische Konzepte in Abfallforschung und -praxis. In: Matschullat J, Müller G (eds) Geowissenschaften und Umwelt. Springer, Berlin Heidelberg New York, pp 315–330
Förstner U (1995a) Non-linear release of metals from aquatic sediments. In: Salomons W, Stigliani WM (eds) Biogeodynamics of pollutants in soils and sediments. Springer, Berlin Heidelberg New York, pp 247–307
Förstner U (1995b) Land contamination by metals – global scope and magnitude of problem. In: Allen HE, Huang CP, Bailey GW, Bowers AR (eds) Metal speciation and contamination of soil. Lewis Publishers, Boca Raton, pp 1–33
Förstner U (1995c) Contaminated aquatic sediments and waste sites: Geochemical engineering solutions. In: Salomons W, Förstner U, Mader P (eds) Heavy metals – problems and solutions. Springer, Berlin Heidelberg New York, pp 237–256
Förstner U, Calmano W, Kienz W (1991) Assessment of long-term metal mobility in heat-processing wastes. Water Air Soil Pollut 57–58:319–328
Förstner U, Kersten M, Wienberg R (1989) Geochemical processes in landfills. In: Baccini P (ed) The landfill – reactor and final storage. Lecture Notes in Earth Sciences. Springer, Berlin Heidelberg New York, 20:39–81
Gambrell RP, Reddy CN, Khalid RA (1983) Characterization of trace and toxic materials in sediments of a lake being restored. J Water Pollut Control Fed 55:1201–1213
Goumans JJJM, Van der Sloot HA, Aalbers ThG (eds)(1991) Waste materials in construction. Studies in Environmental Science 48. Elsevier, Amsterdam, 672 pp
Herman R, Ardekani SA, Ausubel JH (1989) Dematerialization. In: Ausubel JH, Sladovich HE (eds) Technology and environment. National Academy Press Washington DC, pp 50–69
Heron G, Christensen TH (1995) Impact of sediment-bound iron on redox buffering in a landfill leachate polluted aquifer (Vejen, Denmark). Environ Sci Technol 29: 187–192
Hunt CD, Smith DL (1983) Remobilization of metals from polluted marine sediments. Can J Fish Aquat Sci 40:132–142
Kersten M, Förstner U (1991) Geochemical characterization of the potential trace metal mobility in cohesive sediment. Geo-Marine Letts 11:184–187
Kersten M, Moor, C, Johnson CA (1995) Emissionspotential einer Müllver–brennungs–schlacken-Monodeponie für Schwermetalle. Müll und Abfall 11/95, pp 748–758
Kester DR, Ketchum BH, Duedall IW, Park PK (eds)(1983) Wastes in the ocean. Vol 2: Dredged-material disposal in the ocean. Wiley, New York, 299 pp
Lichtensteiger T, Brunner PH, Langmeier M (1988). Klärschlamm in Deponien. EAWAG Project No. 30–681. EC-COST.
Maaß B, Miehlich G, (1988) Die Wirkung des Redoxpotentials auf die Zusammensetzung der Porenlösung in Hafenschlickspülfeldern. Mitt Dtsch Bodenkunde Ges 56:289–294
Malone PG, Jones LW, Larson RJ (1982) Guide to the disposal of chemically stabilized and solidified waste. U.S. Environmental Protection Agency, Washington D.C.
Müller G (1995) Das Schwarze Meer – ein sicheres Endlager für schwermetallkontaminierte Feststoffe?! Geowissenschaften 13:202–206
Neumann-Malkau P (1991) Anthropogenic mass movement – interfering with geologic cycles? In: Geotechnica Congress Cologne, Sept 1991, pp 153–154
Obermann P, Cremer S (1992) Mobilisierung von Schwermetallen in Porenwässern von belasteten Böden und Deponien. Entwicklung eines aussagekräftigen Elutionsverfahrens. Materialien zur Ermittlung und Sanierung von Altlasten. Bd 6. Landesamt für Wasser und Abfall Nordrhein-Westfalen. Düsseldorf, S 127

Pöllmann H (1994) Immobile Fixierung von Schadstoffen in Speichermineralen. In: Matschullat J, Müller G (eds) Geowissenschaften und Umwelt. Springer, Berlin Heidelberg New York, pp 331–340
Roos H-J (1995) Schwermetallentfernung aus Böden und Rückständen mit Hilfe organischer Komplexbildner. Dissertation an der Fakultät für Bauingenieur- und Vermessungswesen der Rheinisch-Westfälischen Technischen Hochschule Aachen, p 221
Salomons W (1993) Non-linear and delayed responses of toxic chemicals in the environment. In: Arendt F, Annokée GJ, Bosman R, van den Brink WJ (eds) Contaminated Soil '93. Kluwer, Dordrecht, pp 225–238
Salomons W (1995) Long-term strategies forhandling contaminated sites and large-scale areas. In: Salomons W, Stigliani WM (eds) Biogeodynamics of pollutants in soils and sediments. Springer, Berlin Heidelberg New York, pp 1–30
Salomons W, Förstner U (Eds)(1988) Environmental management of solid waste: dredged materials and mine tailings. Springer, Berlin Heidelberg New York
Schoer J, Förstner U (1987) Estimation of long-term behavior of heavy metals in sludges. In: Lindberg SE, Hutchinson TC (eds) Proc Int Conf Heavy Metals in the Environment. New Orleans. CEP Consultants, Edinburgh, pp 380–383
Schuiling RD (1990) Geochemical engineering, some thoughts on a new research field. Applied Geochemistry 5: 251–262
Seidel H, Ondruschka J, Stottmeister U (1995) Heavy metal removal from contaminated sediments by bacterial leaching: A case study on the field scale. In: van den Brink WJ, Bosman R, Arendt F (eds) Contaminated soil '95. Kluwer Academic Publ, Dordrecht, pp 1039–1048
Stigliani WM (1992) Chemical time bombs, predicting the unpredictable. In: Chemical time bombs. European state-of-the-art conference on delayed effects of chemicals in soils and sediments. Veldhoven, The Netherlands, Sept 2–5, p 12
Swift RS (1977) Soil organic matter studies. IAEA Vienna, pp 275–281
Turner RR, Lowry P, Levin M, Lindberg SE, Tamura T (1982) Leachability and aqueous speciation of selected trace constituents of coal fly ash. Final Report, Research Project 1061-1/EA-2588. Electric Power Research Institute, Palo Alto, California
Voronkevich SD (1994) Engineering geochemistry: problems and applications. Applied Geochemistry 9:553–559
Wiles CC, Barth E, de Percin P (1988) Status of solidification/stabilization in the United States and factors affecting its use. In: Wolf K, Van den Brink WJ, Colon FJ (eds) Contaminated soil '88. Kluwer Academic Publ, Dordrecht, The Netherlands, 1:947–956
Wilmoth RC, Hubbard SJ, Burckle JO, Martin JF (1991) Production and processing of metals: their disposal and future risks. In: Merian E (ed) Metals and their compounds in the environment. VCH Verlagsgesellschaft Weinheim, Chap I.2, pp 19–65

11 Hydrogeochemical Engineering in Landfills

G. Fleming

11.1 Introduction

A landfill is defined as "*a waste disposal site for the deposit of the waste on to or into land, including internal waste disposal sites (i.e. landfill where a producer of waste is carrying out its own waste disposal at the place of production) and excluding facilities where waste is unloaded in order to permit its preparation for further transport for recovery, treatment or disposal elsewhere, and temporary (i.e. less than one year) deposit of waste prior to recovery, treatment or disposal*" (EEC Draft Directive 1995)

Within Europe, about 240 x 10^6 tonnes of industrial waste, 104 x 10^6 tonnes of municipal waste and more than 60 x 10^6 tonnes of sediment sludge are disposed of each year, of which approximately 65% are landfilled, 24% is incinerated and the balance is partly recycled. European industrial heritage has led to the creation of between 60 000–120 000 landfill sites recorded within the European union which occupy between 800 and 1700 km^2 of land. Only a small proportion of this number are in current operation. Most have been abandoned, restored and the land reused for other purposes. There is no detailed understanding of the environmental impact of these former landfills on water, land and air quality. It is estimated (Boels and Fleming 1993) that all the landfill sites in Europe generate in the order of 12.5 x 10^9 m^3 of landfill gas annually, of which about 755 x 10^6 m^3 have been recovered. The landfill sites also produce between 0.1 and 4.0 x 10^6 m^3 annually of heavily polluted leachate which is dispersed to ground and surface waters. The area of polluted groundwater increases annually by approximately 1–10%, i.e. between 1200 and 12 000 km^2 of contaminated land. Landfill gas contains carcinogenic and pathogenic components, as does the landfill leachate and the gases contribute at least 2% of the total emissions of the greenhouse gases. It must be borne in mind that a main alternative to landfill is to incinerate the solid waste prior to disposing of the ash, however, 30% by volume of the incinerated waste must be landfilled and atmospheric emissions of the incinerators include acids and dioxins.

Waste management in a modern society is complex and essentially involves the use of physical, chemical and biological methods to manage the disposal of solid, liquid and gas wastes. This has led to the evolution of new terminology and concepts including the concept of geochemical engineering (Förstner 1988):

"Geochemical engineering emphasises the increasing efforts of using natural resources available at the disposal site and reducing negative environmental effects of all types of waste material. Future effort should not only be aimed for chemically stabilising critical compounds in wastes materials but in particular for recycling valuable components such as strategic metals".

The wastes and spoils routes for disposal have been identified (Fleming 1992) as shown in Fig. 11.1. Before effective geochemical engineering techniques can be applied to waste disposal, it is important to understand the composition both chemically, physically and biologically of the waste. This is difficult since waste is highly variable from one region to another and from one decade to the next. Figure 11.2 shows a typical trend in waste composition of municipal waste which is an extension of previous work (Crawford and Smith 1985). What the trends clearly show are that there is a shift over the last 50 years away from burning fossil fuels for domestic heating with the associated decline in the ash content of domestic refuse. At the same time there is a large increase in the use of paper in society and an increasing trend in glass and plastic, both of which have now levelled off. The difficulty for a landfill designer or a waste disposal manager is to be able to predict the future trends in composition of waste. This must take into account the current legislation to encourage recycling, while at the same time assess change in manufacturing and consumer demand which has led to a large increase in the generation of electronic equipment. The European Commission estimates there are now 6 million tonnes/year of "end-of-life" electronic equipment (Times 1995). This in time will lead to a further increase in plastic, metals and glass appearing in the waste stream from such sources. The landfill designer needs to ensure that the waste stream is appropriate for the disposal to a specific landfill design i.e. dilute and disperse or containment. Hence, there is a need to engineer the waste stream to suite the design. This is not an option practised by landfill designers.

Present options in the disposal route vary throughout Europe and Table 11.1 (see p. 187; National Geographic 1991) shows the waste disposal routes preferred by various countries in Europe and the United States. From this it is clear that landfill is still the favoured option in waste disposal, although in recent years particularly in Germany and Denmark, a major push is taking place towards incineration. It must be remembered (Förstner 1988) that the reduction in the organic fractions of municipal waste by simple incineration still leaves 30% by volume of ash to be disposed of, and this ash contains much higher concentrations of toxic pollutants including chlorides and heavy metals which it is argued (Förstner 1988) can be reutilised if geochemical engineering methods are applied to the waste ash.

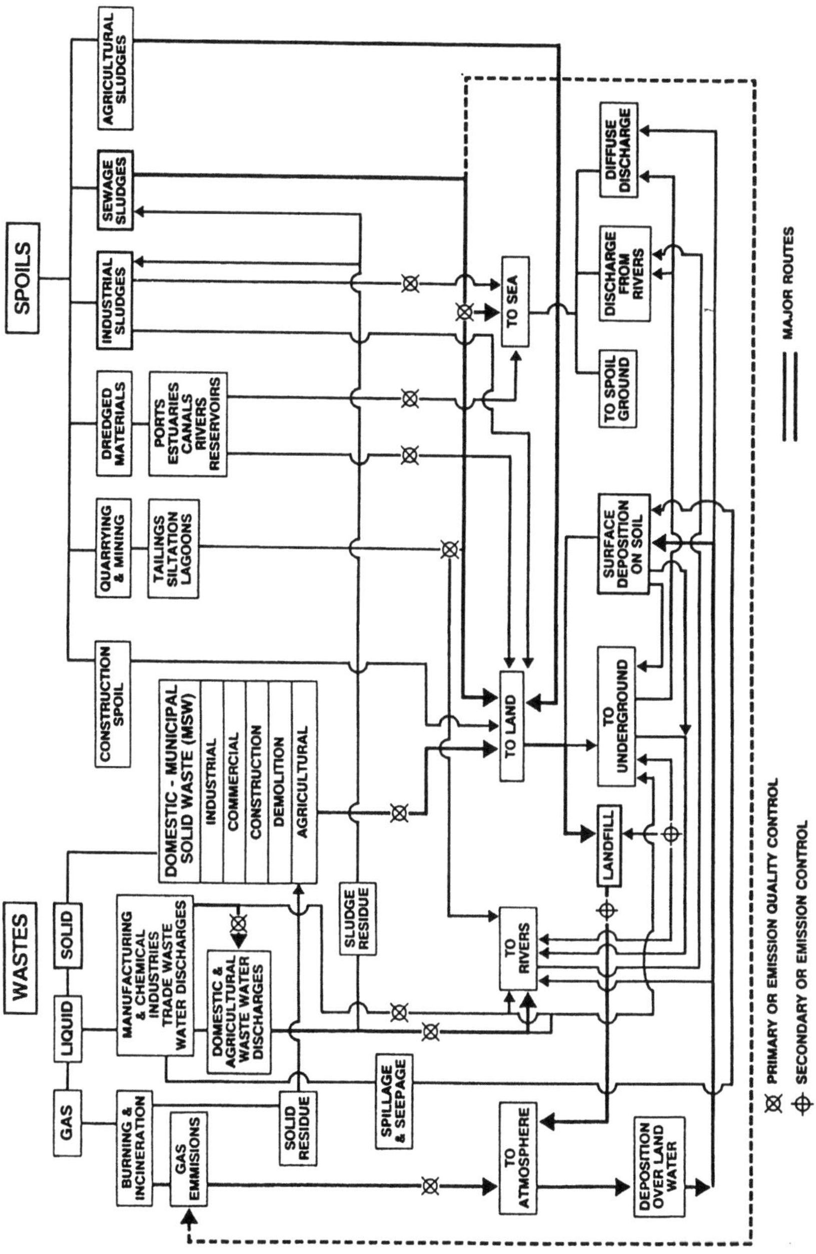

Fig. 11.1. Waste and spoils and routes for disposal (Fleming 1992)

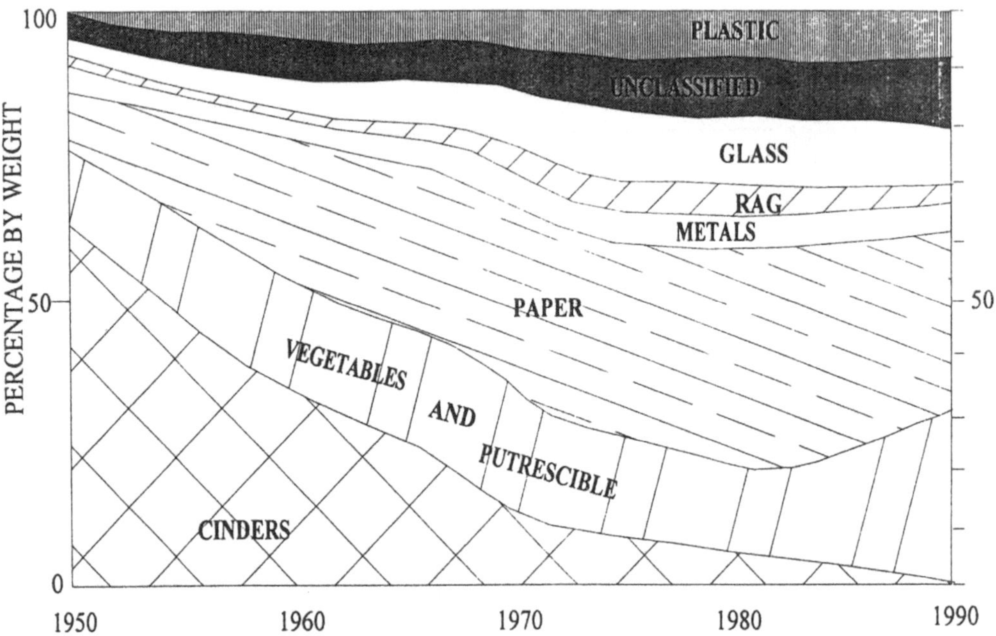

Fig. 11.2. What goes into a typical municipal waste landfill? (Crawford and Smith 1985; Up-dated 1995)

The problem of waste disposal is compounded by the amount and complexity of current European legislation, e.g. the disposal of sewage sludges to sea will cease in 1997 and result in this route for disposal being closed. This will place a greater emphasis on alternative disposal routes including land disposal, incineration or landfill. In addition the EEC draft legislation on landfill (EEC Draft Directive 1995) places considerable pressure on a move towards containment landfill and the concept of mono-landfill. The difference between mono-landfill and co-disposal landfill is that in a mono-landfill, only one waste type is accepted, and this logically will be either a sludge or an ash and involves only inorganic wastes. Mono-landfill is therefore the separation of waste into inorganic/organic fractions. Co-disposal involves the disposal of a mixture of organic and inorganic materials, and results in what is often referred to as the reactor landfill. In the case of mono-disposal landfills, the process is dominantly physical or chemical, whereas, in the reactor landfill the physical, chemical and biological processes are interactive. Within the UK recent legislation including the Environmental Protection Act (Environmental Protection Act 1990) has led to stringent laws and rules controlling every aspect of waste, collection, transportation, disposal and after-care. Irrespective of legislation currently in force and

Table 11.1. Routes for (%) waste disposal (Fleming 1992)

Country	Landfilled	Incinerated	Recycled	Composted
Austria	64	20	–	16
Denmark	31	50	18	1.0
France	47.9	41.9	0.6	8.7
Germany	74	24	–	2
Italy	83.2	13.9	0.6	2.3
The Netherlands	51	34	15	–
Sweden	35	60	5	–
Switzerland	20	80	–	–
United Kingdom	88	11	1.0	–
United States[a]	70	30		

[a] National Geographic 1991

the proposed EEC directive on landfill, it is clear landfill will play a very important part in future waste management strategy. It is therefore vital to ensure that landfills are engineered to ensure that they do not impact on the environment in a detrimental way. Future engineering of landfills will most certainly involve hydrogeochemical engineering, as well as engineering the waste stream to recycle the valuable primary products.

11.2 Landfill Processes

This section reviews the reactor landfill and the processes present within the landfill system. The primary objective of a landfill system is to ensure that waste deposited within the system reaches a final storage quality as rapidly as possible which matches background levels of the earth's crust. Past landfill strategy has simply been to deposit waste in derelict voids or land depressions in the most cost-effective manner with no consideration of the long-term impact on the environment.

On one hand, the waste mix contains a range of contaminants which when co-disposed in a reactor landfill, interact both physically, chemically and biologically and in turn cause a change to the state of the waste by converting the solid material into various liquid, gases and residues. The emissions from landfill sites are to air and water and the final residue reaches a final storage quality after decades or

Table 11.2. Typical leachate composition from landfill sites (Christianson et al. 1994)

Parameter	Range
pH	4.5–9
Spec. cond. (μS/cm)	2500–25,000
Total solids	2000–60,000
Organic Matter	
Total organic carbon (TOC)	30–27,700
Biological oxygen demand (BOD_5)	20–57,000
Chemical oxygen demand (COD)	140–90,000
BOD_5/COD (ratio)	0.02–0.80
Organic nitrogen	14–2500
Inorganic Macrocomponents	
Total phosphorous	0.1–20
Chloride	150–4500
Sulfate	8–7750
Hydrogencarbonate	610–7320
Sodium	70–7700
Potassium	50–3700
Ammonium-N	50–1800
Calcium	10–7200
Magnesium	30–15,000
Iron	3–5500
Manganese	0.03–1400
Heavy Metals	
Arsenic	0.01–1
Cadmium	0.0001–0.4
Chromium	0.02–1.5
Cobalt	0.005–1.5
Copper	0.005–10
Lead	0.001–5
Mercury	0.00005–0.16
Nickel	0.015–13
Zinc	0.03–1000
Bacteria	
Fecal streptococci (CFU ml^{-1})[a]	0.1–3 million
Fecal coliform (CFU ml^{-1})	0.1–100,000
Total coliform (CFU ml^{-1})	0.1–100,000

Note: The ranges are based on Andreottola and Cannas (1992), Ehrig (1980), (1983), (1988), Garland and Mosher (1975), Johansen and Carlson (1976), Karstensen (1989), Lu et al. (1985), Naturvårdsverket (1989), Robinson and Maris (1979). Values in milligrams per liter unless stated otherwise.

[a] CFU, colony forming units.

centuries. Gaseous and liquid emissions from landfill contain pollutants derived from the waste mix which are hazardous to the environment. These pollutants can be divided into gaseous and liquid emissions. Typical leachate composition from landfill sites is shown in Table 11.2 (Christianson et al. 1994) which when broken down into the different phases of landfill from the acid to the methanogenic phase have typical values shown in Table 11.3 (Ehrig 1983, 1988). The contaminants to leachate can be generally broken down to four groupings of pollutants:

1. Dissolved organic matter.
2. Anthropogenic specific organic compounds (Christianson et al. 1994).
3. Inorganic macrocomponents.
4. Heavy metals.

Table 11.3. Leachate composition in terms of average values and ranges for parameters with differences between acid and methanogenic phase (Ehrig 1988) and average values for parameters with no observed differences between acid and methanogenic phase (Ehrig 1983)

	Acid phase		Methanogenic phase		
Parameter	Average	Range	Average	Range	Average
pH	6.1	4.5–7.5	8	7.5–9	
Biological oxygen demand (BOD_5)	13,000	4000–40,000	180	20–550	
Chemical oxygen demand (COD)	22,000	6000–60,000	3000	500–4500	
BOD_5/COD (ratio)	0.58		0.06		
Sulfate	500	70–1750	80	10–420	
Calcium	1,200	10–2500	60	20–600	
Magnesium	470	50–1150	180	40–350	
Iron	780	20–2100	15	3–280	
Manganese	25	0.3–65	0.7	0.03–45	
Ammonia-N					741
Chloride					2120
Potassium					1085
Sodium					1340
Total phosphorus					6
Cadmium					0.005
Chromium					0.28
Cobalt					0.05
Copper					0.065
Lead					0.09
Nickel					0.17
Zinc	5	0.1–120	0.6	0.03–4	

Note: All values in milligrams per liter except pH and BOD_5/COD.

Table 11.4. The most frequently observed anthropogenic-specific organic compounds (ASOCs) in landfill leachates (Christianson et al. 1994)

Compound	Range (µg/l)
Aromatic Hydrocarbons	
Benzene	1–570
Toluene	1–7500
Xylenes	4–3500
Ethylbenzene	1–1100
Trimethylbenzenes	4–250
Naphthalene	0.1–260
Diethylphthalate	10–660
Di-*n*-butylphthalate	5.0–15
Butylbenzylphthalate	5.1–8
Halogenated Hydrocarbons	
Chlorobenzene	0.1–110
1,2-Dichlorobenzene	0.1–32
1,4-Dichlorobenzene	0.1–16
1,1-Dichloroethane	0.6–46
1,2-Dichloroethane	<6
1,1,1-Trichloroethane	0.1–90
trans-1,2-Dichloroethylene	1.6–88
cis-1,2-Dichloroethylene	1.4–470
Trichloroethylene	0.7–750
Tetrachloroethylene	0.1–250
Methylene chloride	1.0–64
Chloroform	1.0–70
Carbontetrachloride	4.0–9.0
Phenols	
Phenol	1–1200
Ethylphenols	<300
Cresols	1–2100
Pesticides	
Mecoprop[2]	2.0–90
2,4-D[3]	1.0–5
Miscellaneous	
Acetone	6–4400
Tetrahydrofuran	9–430
Methylethylketone	110–6600
Tri-*n*-butylphosphate	1.2–360
Triethylphosphate	15
Camphor[4]	1[5]
Fenchone	20–34

The anthropogenic specific organic compounds present in landfill leachates have been listed (Christianson 1994) and are shown in Table 11.4. When considering the gas composition of emissions from landfill, typical results are shown in Table 11.5 (HMSO 1989).

The objective of landfill management is to define a landfill system which controls the biological breakdown of the organic element of the waste mass, and can result in the emission streams to air and water which can be further controlled to ensure that they are environmentally friendly. In order to understand the ability to control a landfill system, it is important to understand the processes within the landfill system. Figure 11.3 (Fleming 1992) shows the landfill starting conditions which are crucial to the understanding of the landfill processes. From this system it can be clearly identified that there is a number of capacity-controlling characteristics within a landfill which will determine the final quality of the landfill gas and leachate emissions. These capacity-controlling characteristics include the waste structure, the cation exchange capacity, the total organic carbon, the pH, the redox

Table 11.5. Typical landfill gas composition (HMSO 1989)

Component	Typical value (% Volume)	Observed maximum (% Volume)
Methane	63.8	88.0
Carbon dioxide	33.6	89.3
Oxygen	0.16	20.9
Nitrogen	2.4	87.0
Hydrogen	0.05	21.1
Carbon monoxide	0.001	0.09
Ethane	0.005	0.0139
Ethane	0.018	–
Acetaldehyde	0.005	–
Propane	0.002	0.0171
Butanes	0.003	0.023
Helium	0.00005	–
Higher alkanes	<0.05	0.07
Unsaturated hydrocarbons	0.009	0.048
Halogenated compounds	0.00002	0.032
Hydrogen sulphide	0.00002	35.0
Organosulphur compounds	0.00001	0.028
Alcohols	0.00001	0.127
Others	0.00005	0.023

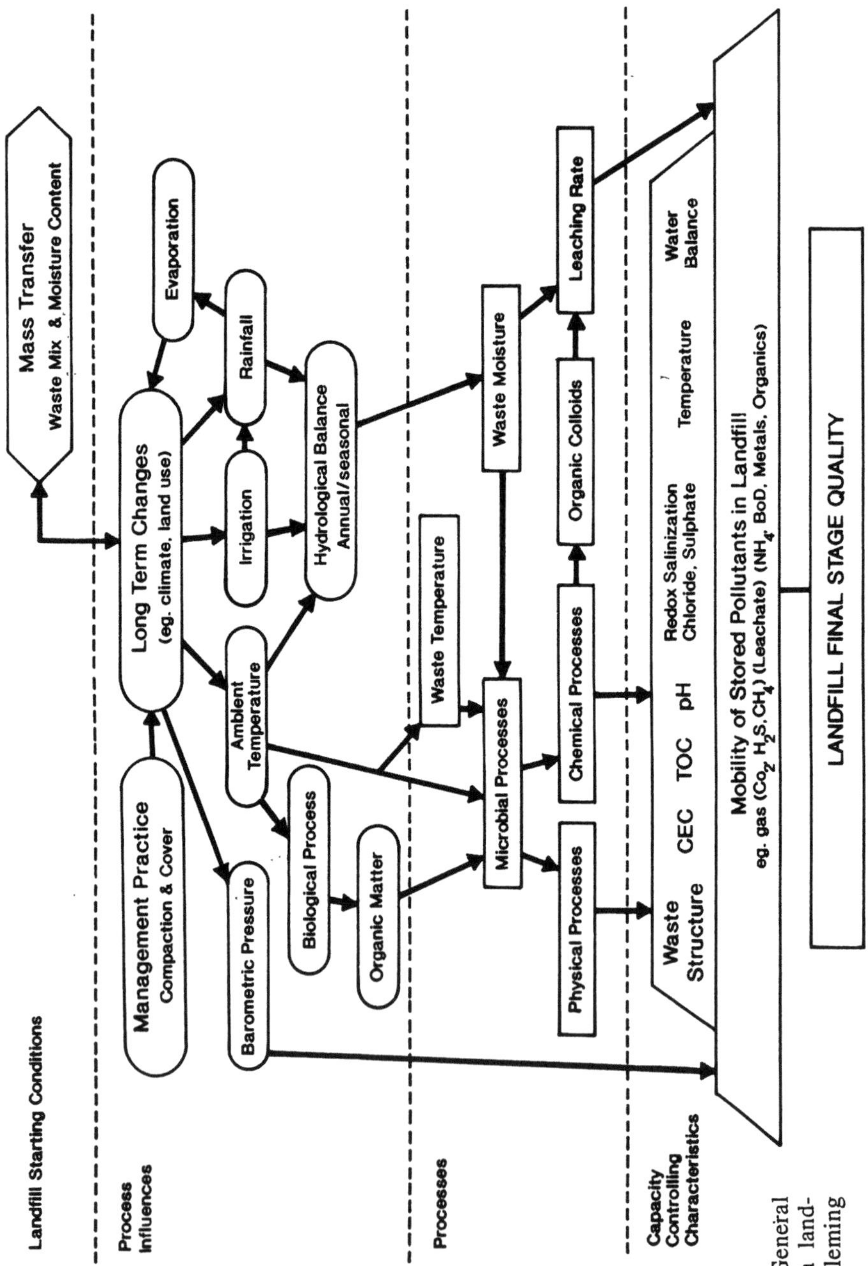

Fig. 11.3. General processes in a landfill system (Fleming 1992)

and salinisation potential, the temperature of the landfill site and its water balance. In the design and operational phases of landfill, it is crucial to achieve the optimum conditions of each of the capacity-controlling characteristics in order to optimise the rapid development of the landfill process to reach a final stage quality in a reasonable period of time. It is generally understood that a number of the capacity-controlling factors include a pH of between 6.5 and 8.5, a waste moisture content between 50 and 85%, a moderate to high density of waste and a temperature of between 35° and 45°C. Naturally each capacity-controlling characteristic depends on the waste composition which in turn dictates the chemical and biological characteristics of the waste mix. The most important fact to remember in all landfill engineering is that every landfill is site specific and the design and operation of a site must be based on the conditions and waste stream typical to an individual site. This in turn is influenced by the fact that the climate is site specific and time dependant. Waste is highly variable both in time and in physical, chemical and biological composition. In addition the rock structure and the hydrogeology are unique for each site.

11.3 Siting a Landfill

One of the most difficult steps in developing a waste disposal strategy is to identify suitable sites for landfill operations. Within the European context in a highly urbanised community, sites suitable for landfill are scarce. The process of choosing a landfill site is subject to a complex regional and local planning process which embraces a large number of physical and political decisions. The situation in North America (Noble 1992) is similar to that in Europe, in that a detailed procedure has to be set in motion before a landfill can be legally sited at a specific location. Table 11.6 shows the landfill siting criteria which is generally applicable, and involves initially a regional assessment based on land use and natural features to determine areas which are suitable. The first phase in siting a landfill involves a primary sieve analysis where the existing land use and natural features are superimposed one on top of the other using a colour coding such that areas suitable for landfill are identified on a regional basis. The second stage of siting a landfill is to consider local criterion which involves a secondary sieve involving initially a full environmental impact assessment of the proposed sites within a given region, taking into account all aspects of environmental impact from surface and groundwaters through historical and archaeological sites, restrictions on highways, traffic movements and the proximity of the landfill site to the centre of waste generation. The secondary sieve analysis is then undertaken where each site is ranked on the basis of the impact for each category of environmental impact and the sites satisfying the siting criteria are chosen. A preliminary design phase is then undertaken to determine the method of waste disposal, taking into account the

Table 11.6. Siting criteria for landfill development (Noble 1992)

Natural Features
Wetlands
Flood plains
Surface waters
Groundwater resources
Fault zones
Seismic impact zones
Unstable areas
Expansive soils
Subsidence zones
Threatened and endangered species habitat
Landfill Resources
Sufficient depth of suitable soils for separation from groundwater resources
Sufficient depth of suitable soils for cover material
Existing depressions
Natural screening
Limited run on potential
Significant depth to groundwater resources
Minimal residential well density
Absence of groundwater conditions which make monitoring difficult
Minimal slope
Land Use
Development (existing or committed)
Airports (jet or turboprop)
Prime farmland
Municipal water supply wells
Areas of historic importance
Areas of architectural importance
Areas of paleontological importance
Areas of archaeological importance
Areas with natural buffering
Proximity to areas with final use compatibility
Proximity to municipal boundaries
Areas of natural scenic beauty
Economic Factors
Proximity to major highways
Highway restrictions
Traffic
Distance from centre of waste generation or transfer
Availability
Land holding in large parcels

availability of natural materials. An application for planning permission is made at which all the evidence and information leading to the final site selection is presented together with a detailed design involving an engineering plan and details of the hydrogeochemical management strategy for the site, including after-care and site monitoring.

11.4 Landfill Design

Landfill design requires the following steps (Rowe et al. 1995):

1. Assess the hydraulic conductivity of aquitards in the context of the proposed engineering design;
2. Recognise that for highly engineered systems the hydraulic conductivity which will give the greatest potential impact is not necessarily the highest or lowest reasonable value, but may lie in between reasonable bounds;
3. Give consideration to the potential effect of the proposed facility on groundwater levels and long-term performance of the landfill system (i.e. the shadow effect);
4. Give consideration to the service life of each of the engineered components of the barrier system (primary and secondary leachate collection, liners, hydraulic control layers etc.) with respect to the contaminating life-span of the landfill;
5. Design a system such that there will be negligible impact on groundwater even after the service lives of the engineered components of the barrier system have been reached;
6. Monitor the performance of the system and match monitored performance against that expected. "Trigger" conditions may be developed (e.g. level of leachate mounding and concentration of contaminants in the leachate; concentrations at volume of fluid collected from secondary collection systems) such that back-up measures can be implemented once a failure of part of the system has been detected, and before conditions have been established which will ultimately cause contamination of groundwater.

The detailed design of the landfill depends very much on the criteria discussed in siting a landfill in the first place. Historically landfills have been operated as a dilute and disperse systems (Fig. 11.4) or as a containment system (Fig. 11.5). The majority of past landfills have been designed on the dilute and disperse principal by accident or default and a significant number continue in operation in the present time with many operating satisfactory. With the current EEC Directive on Landfill, a shift has taken place towards the principal of containment where both gas and liquid emissions from a landfill are controlled in such a manner that both

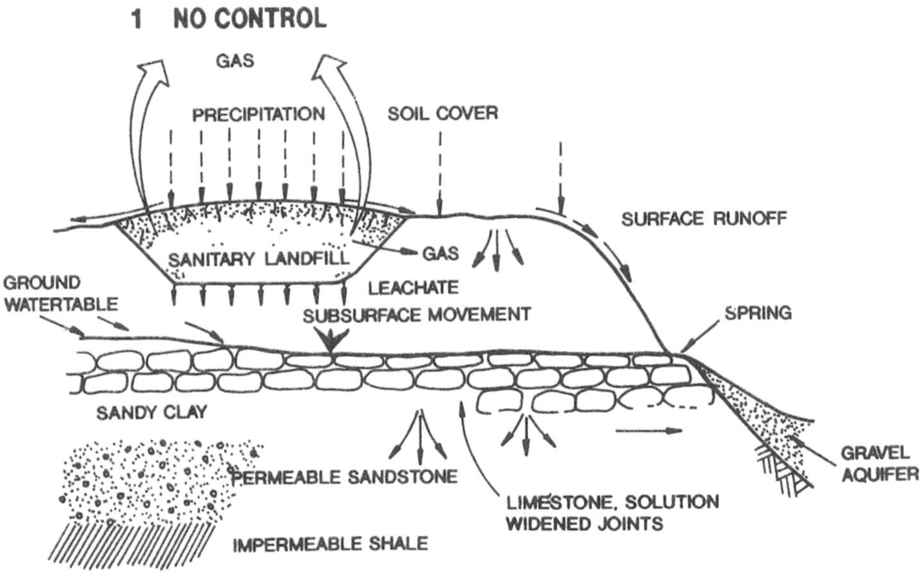

Fig. 11.4. Active or closed landfill (Webster 1988)

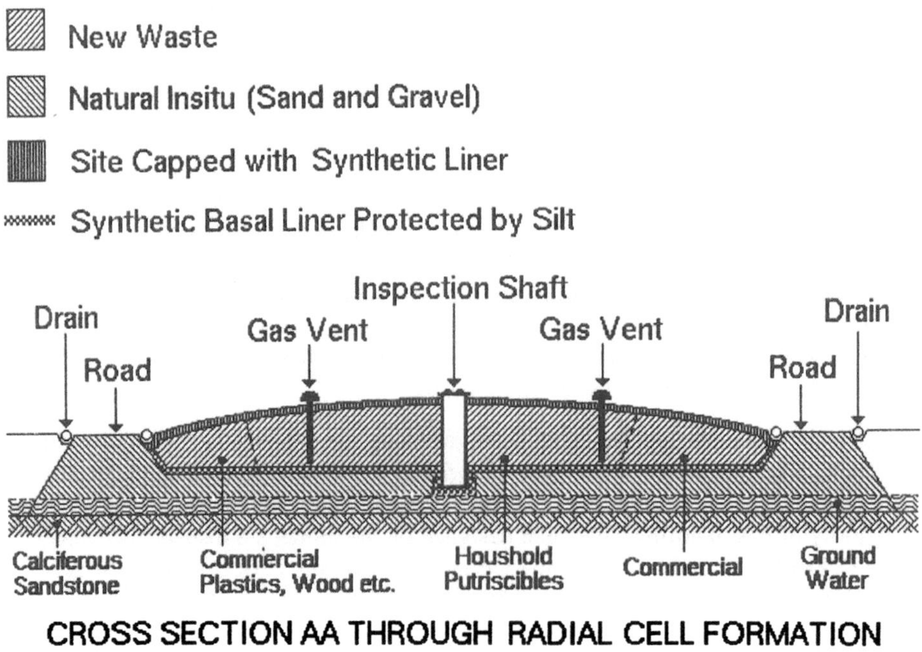

Fig. 11.5. Closed landfills – containment

the gas and liquid emissions can be treated before dispersal into the atmosphere or into the surface water and groundwater systems. The dilute and disperse principal allows the free entry of rainfall and liquids into the landfill and allows a natural dispersion of the leachate generated from a landfill into the rock structure and groundwater system beneath and adjacent to the landfill. A containment system is based on the principal of designing a basal liner to the landfill site which controls the rate of which emissions take place through the landfill site base and involves also a cap to the landfill site which controls the gaseous emissions to the atmosphere. Both the basal liner and landfill cap can be designed from natural materials of known permeability, such as natural clays or reworked clays, or from engineered physical barriers of known permeability, including cut-off walls, natural bedrock or geomembranes in composite liner systems.

It is likely that over the next decade engineering of new landfill sites will tend to favour the containment approach but management of existing and new landfill sites should take into account the impact of existing dilute and disperse landfills on the groundwater and atmospheric pollution. During this transition a number of landfills based on the dilute and disperse principal have failed and the redesign of these landfill sites often involves engineering, basal liners or landfill caps to control the process. When converting a landfill from a dilute and disperse principal to a partial dilute and disperse principal or a full containment system, it must be stressed that no landfill site can be engineered to be 100% impermeable.

11.5 Landfill: Transport Mechanisms

It is important to understand that when designing a landfill site to control the transport of contaminants through saturated clay barriers or engineered membranes, an understanding of the primary transport mechanisms is essential. These primary transport mechanisms include *advection*, which involves the movement of contaminants with flowing water, and *diffusion*, which involves the movement of contaminants from points of high chemical concentration to points of low chemical concentration. Once the contaminant is transported out of the landfill site, a third transport mechanism may come into play involving the process of *dispersion*, which consists of mixing which occurs due to local variations in the flow velocity of the groundwater. This mixing causes dilution. In addition to the transport mechanisms which allow the contaminants to move in the groundwater, it is also important to note that the process of *sorption* where contaminants are removed from solution by cation exchange whereby cations such as sodium, lead, cadmium, iron and copper replace other cations such as calcium and magnesium on the surface of the clay barrier, and hence are removed from solution. However, this usually leads to higher concentrations of the replaced cations, e.g. calcium and magnesium in the fluid concentration. In addition to sorption, the process of

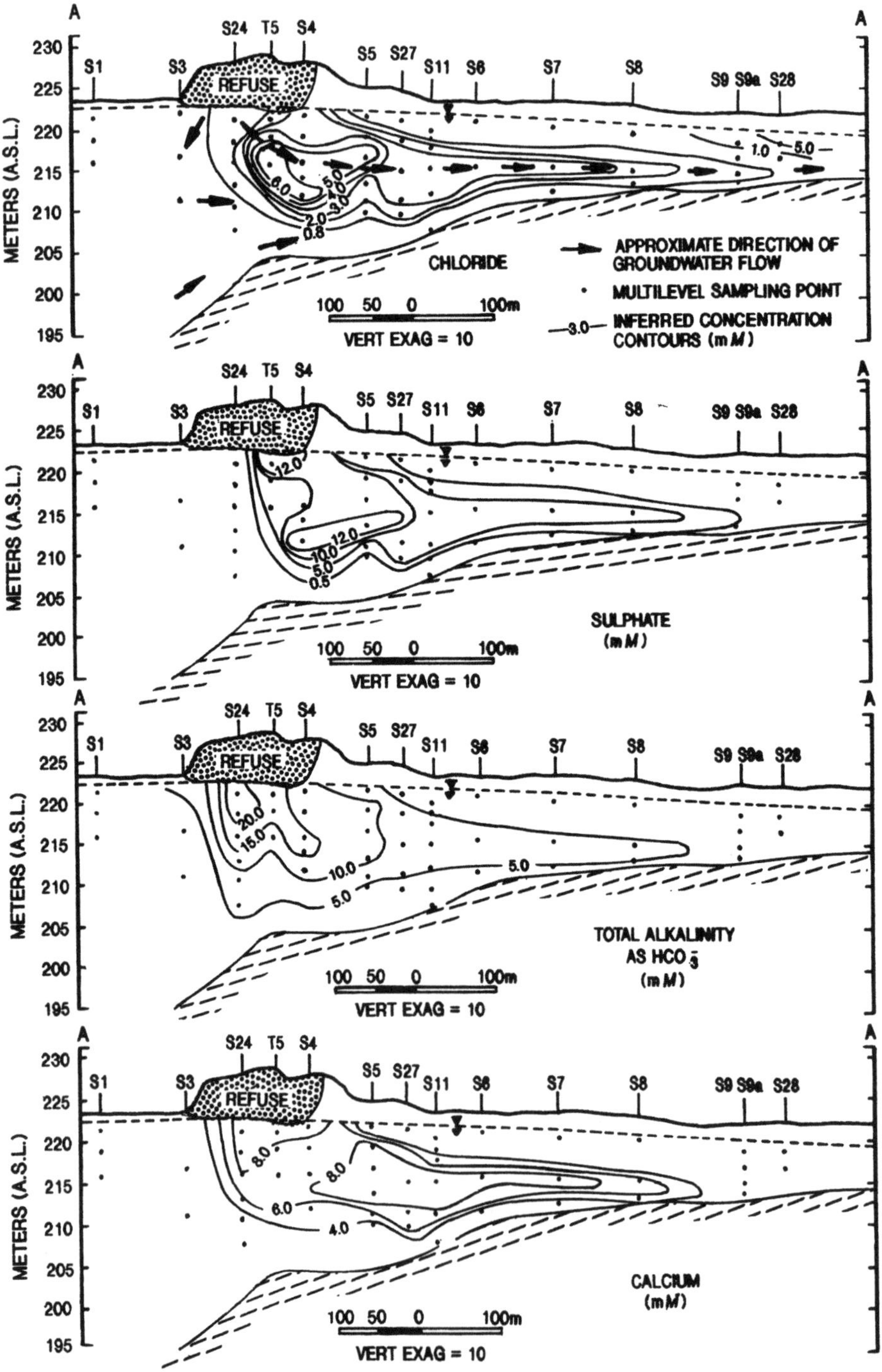

Fig. 11.6 Concentration contour plots of the Borden Landfill, Canada, leachate plume with respect to chloride, sulphate, alkalinity (HCO_3-) and calcium (Nicholson et al. 1983)

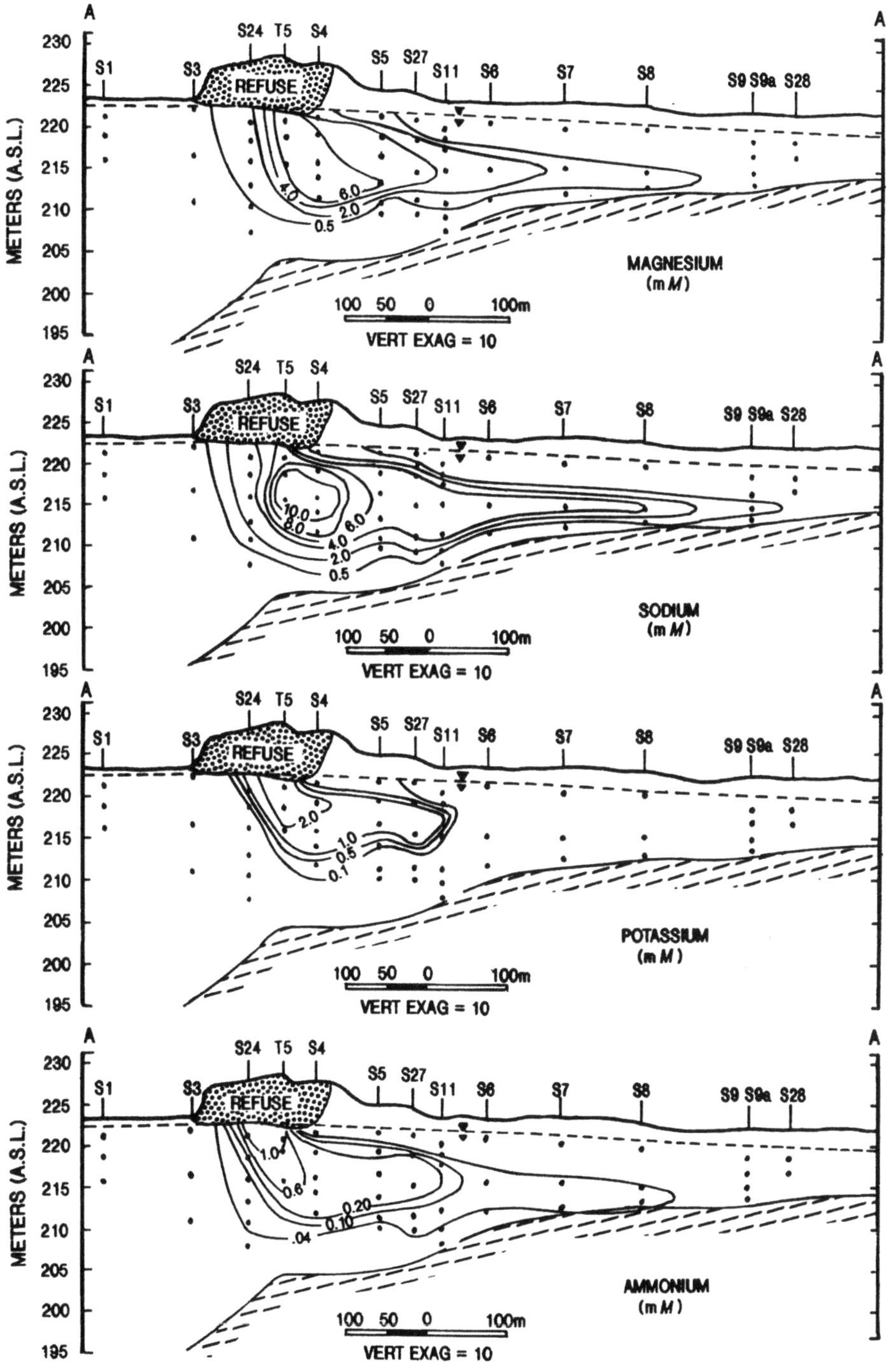

Fig. 11.6 continued Concentration contour plots of the Borden Landfill leachate plume with respect to magnesium, sodium, potassium and ammonium (Nicholson et al. 1983)

precipitation can take place where heavy metals are deposited from the solution as a solid. A further process which can be included in the processes covered by sorption is the removal of organic contaminants from solution by the interaction with solid organic matter in the soils or rocks. Figure 11.6 shows the movement of contaminants from a landfill site under a dilute and disperse process. It should be noted that the processes of dispersion and sorption will be taking place.

In the design of a landfill system, it is possible to have a combined advective diffusive transport mechanism such as is shown in Fig. 11.7 (Rowe et al. 1995). Figure 11.7 shows a natural clay deposit and a downward advective diffusive transport mechanism, whereas if the landfill site is constructed within the potential water surface of the aquifer, then within the clay deposit upward advection can be achieved with no migration of pollutants from the landfill site by advection, but downward diffusion can still take place. It must be stressed that for the design of landfill sites, no matter what geomembrane or natural clay barrier is used, there will still be leakage through the geomembrane liner. Table 11.7 (Giroud and Bonaparte 1989) shows the leakage rates through geomembranes (Giroud and Bonapart 1989).

Natural clay barriers appear to be the best approach to engineering landfill sites, assuming that the natural clays have appropriate permeabilities and if they are reworked natural clays, they have to be reworked within a strict range of moisture content to achieve the necessary permeability. Other factors which help improve the performance of clay liners with time include the replacement of calcium and magnesium absorbed on most natural clays by the sodium in landfill

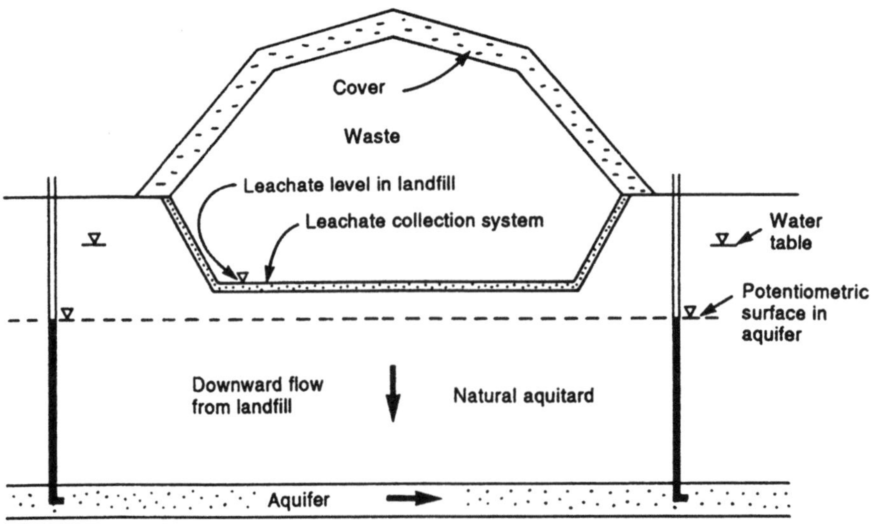

Fig. 11.7. Barrier design involving a leachate collection system, a natural clayey deposit and downward advective-diffusive transport (Rowe et al. 1995)

Table 11.7. Estimates of leakage through geomembrane liners (Giroud and Bonaparte 1989)

Type of liner	*Leakage mechanism*	*Leakage rate (l/ha per day) for a liquid depth on top of the geomembrane, h (m) of*				
		0.003	*0.03*	*0.3*	*3*	*30*
Geomembrane alone (between two pervious media)	Permeation	0.0001	0.01	1	100	300
	Small hole	100	300	1 000	3 000	10 000
	Large hole	3000	10 000	30 000	100 000	300 000
Composite liner (good field conditions)	Permeation	0.0001	0.01	1	100	300
	Small hole	0.02	0.15	1	9	75
	Large hole	0.02	0.2	1.5	11	85
Composite liner (poor field conditions)	Permeation	0.0001	0.01	1	100	300
	Small hole	0.1	0.8	6	50	400
	Large hole	0.1	1	7	60	500

The large hole has a surface area of 1 cm^2. The frequency of holes is 1 per 4000 m^2. The thickness of the soil layer is 0.9 m and its hydraulic conductivity is 10^{-9} m/s. The HDPE geomembrane thickness is 1 mm. Leakage rates in the case of a composite liner do not significantly depend on the material overlying the geomembrane. In the case of a geomembrane alone, leakage rates were calculated assuming that the geomembrane is overlain and underlain by an infinitely pervious membrane. This assumption is valid for coarse gravel or geonet. Leakage rates through holes would be significantly less if the geomembrane were overlain and/or underlain by sand or a less permeable material.

leachate (Rowe et al. 1995). Also according to research (Peters 1993), the depth of clay barrier effected by chemical change is relatively small and usually only limited to a few centimetres over a number of decades. The use of capillary barriers as landfill caps is yet another potential tool in landfill design

In designing landfill sites or redesigning existing landfill sites to modern standard, it is necessary also to take into account the effectiveness of the landfill cap. In order to demonstrate this, a case study is considered at the Brackletter landfill site in the west of Scotland (Fleming 1989). This landfill site was engineered as a dilute and disperse landfill and the design calculations demonstrated that the expected contaminant transport from the landfill would be diluted within the attenuation zone before reaching the nearby water course. In effect, when the landfill site was constructed, the attenuation zone failed to provide sufficient dispersion and diffusion capability and the discharge to the nearby water courses became polluted as shown in Fig. 11.8. One of the principal engineering solutions to remediate the site was to cap the site with a one-millimetre-high density polyethylene membrane to prevent the ingress of high surface rainfall in the order of 1.5 metres per year in excess of evaporation. The result of engineering with the geomembrane effectively reduced the pollutant flow from the landfill site in a relatively short time (see Fig. 11.9).

There is little doubt that with modern engineering knowledge of landfills, it is possible to engineer a landfill to not produce leachate and to effectively control any temporary leachate generated during the early stages of landfill infilling. It is also feasible to intercept all the gases from a landfill site and to treat these before discharge to the atmosphere.

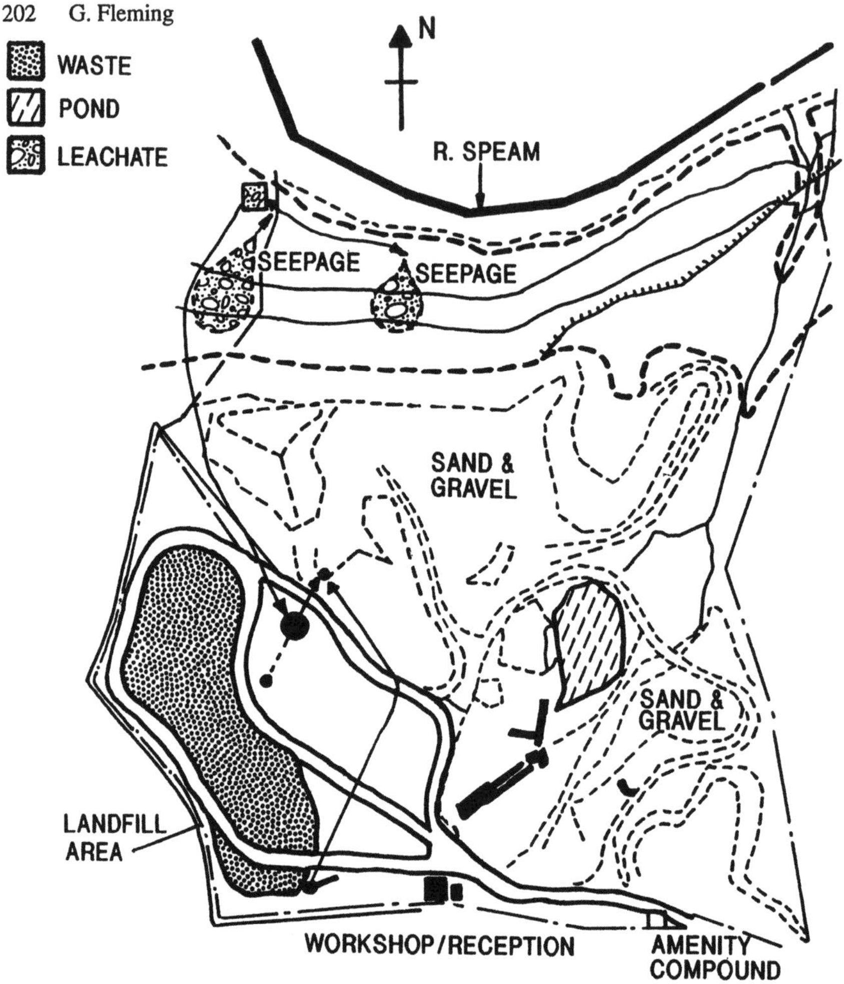

Fig. 11.8. Remedial works at Brackletter (Fleming 1989)

An example of a successful operation of dilute and disperse landfill site is one located at Greenoakhill, near Glasgow, Scotland (Fleming 1991). This landfill site was initated in the late 1950s and is operated continuously to the present time with an expected 20-year life beyond 1995. The site infilled a mixture of industrial wastes, commercial wastes, household waste and special liquid wastes, including sludges. In parallel, the site was operated for the extraction of sand and gravel. Excavation of sand and gravel preceeded the waste disposal. Continuous monitoring of the site including work in 1976 (Harrison 1976; Fig. 11.10) has shown that the leachate plume resulting from the disposal of this complex amount of wastes (now in excess of 10 million m^3) has not migrated outwith the boundary of the site and has not resulted in serious pollution of the underlying groundwaters above that which has historically occurred due to mining and other activities.

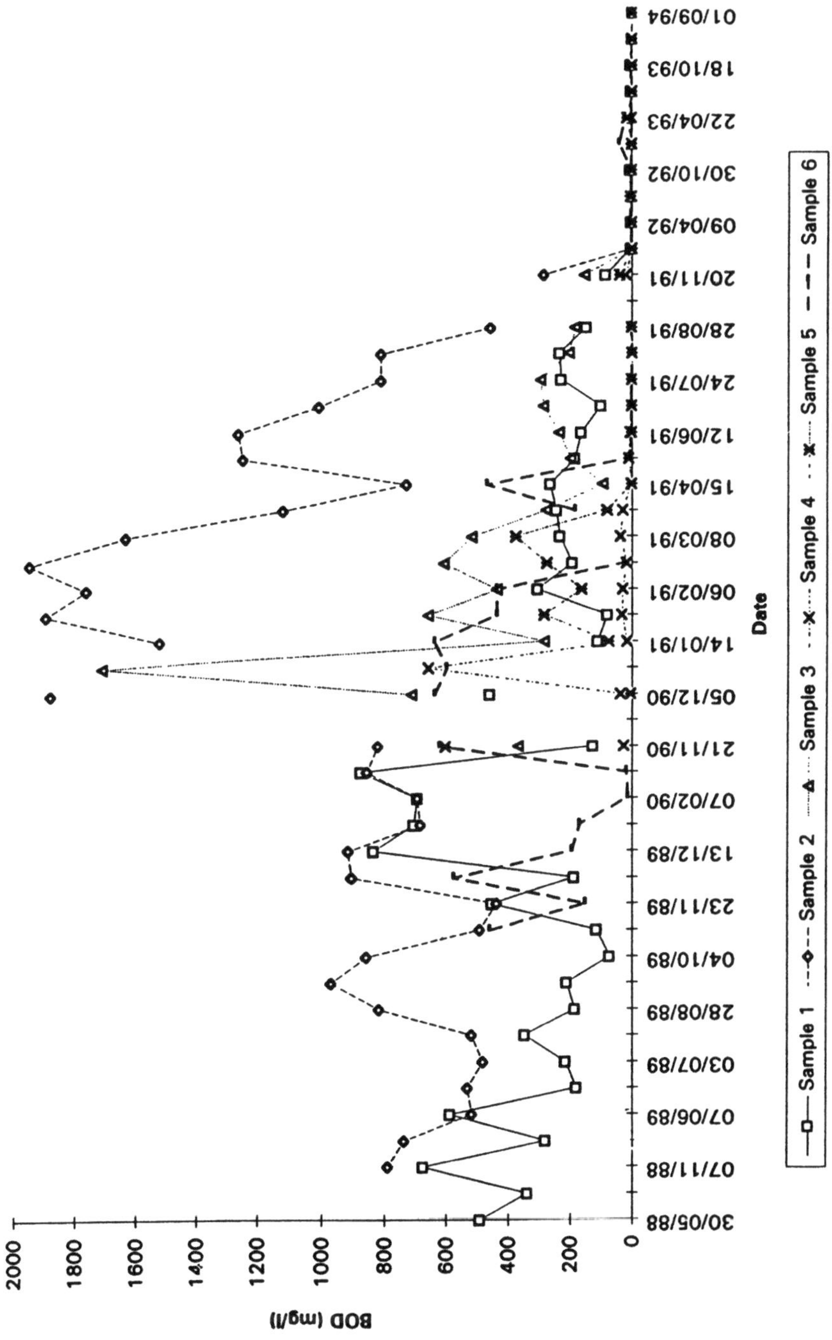

Fig. 11.9. Brackletter leachate analysis (Fleming 1994)

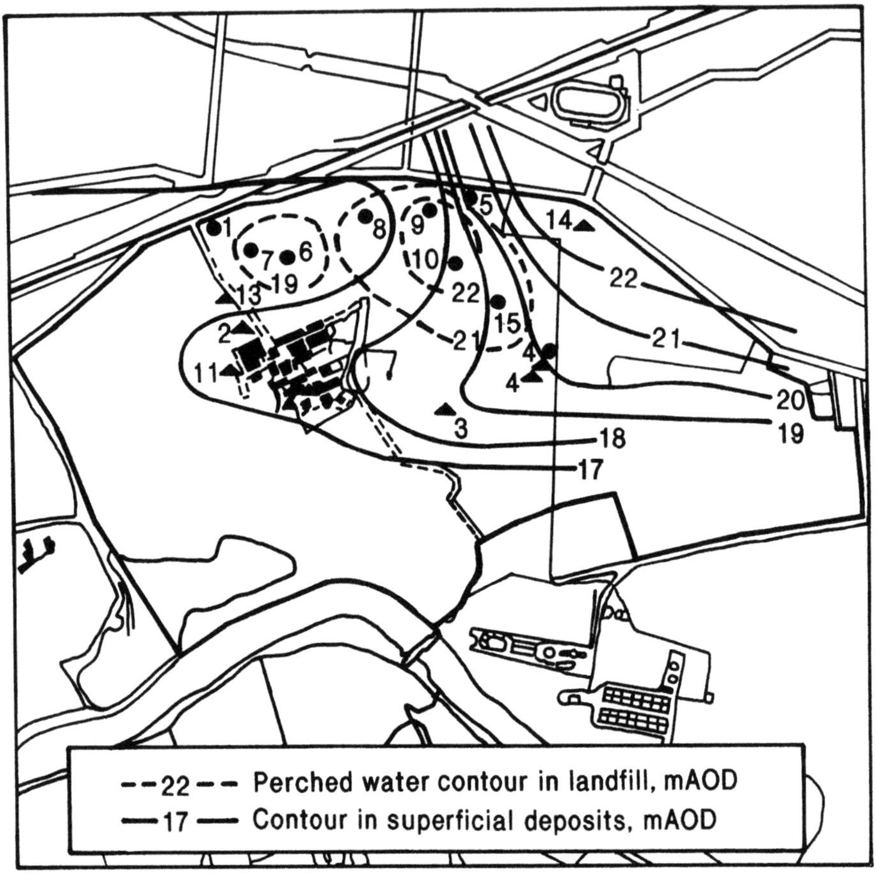

Fig. 11.10. Leachate movement according to Harrison's Report (1976)

11.6 Landfill Operational Management

The design of a landfill site is only one part of the process of ensuring that landfills do not impact adversely on the environment. The second important part of good landfill practice is the operational management of the landfill to ensure that the waste stream is controlled, that recycling of key natural resources is optimised and that the final residue deposited on the landfill site is at the optimum conditions for rapid biodegradation to a final stage quality which is environmentally compatible.

The growing shortage of natural mineral resources and the trend in recycling construction spoils of concrete, brick and soils is accelerating. At the same time the process of recycling paper, plastic, glass and metal from the waste stream is

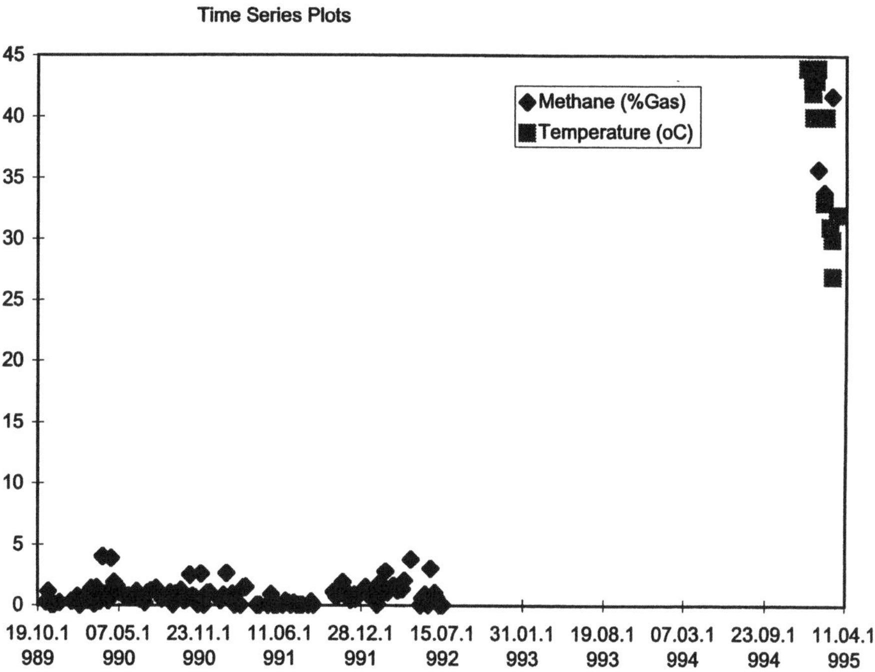

Fig. 11.11. Craigmuschat landfill site – transition into methanogenesis with accompanying rise in temperature

also accelerating with hopefully between 25 and 50% recycling by weight being achieved by the year 2000. The impact of this on the final waste residue to be disposed in a landfill site has to be assessed. It is interesting to note that recent research (Jones 1994) indicates that the removal of plastic and glass from the waste disposal stream can seriously reduce the surface areas for bacteria to breed on and react with the biodegradable waste, hence, the removal of plastic and glass can slow down the biodegradation process. It is also interesting to note that recent research (Fleming and Macdonald 1994) has shown that if the moisture content of the landfill site is not maintained at the optimum, then biodegradation may not take place at all.

In a recent case study for the Craigmuschat site in the west of Scotland (Fleming 1989), at that site infilling of medium density baled waste in a well-drained environment resulted in no significant biodegradation over a six-year period, however, once the site was sealed and the ingress of oxygen limited and the temperature of the site increased, then methanogenesis was achieved. An interesting feature of this is that the temperature profile between the surface and the base of the site was considerable, the temperatures optimum to methanogenesis at the base of the site were satisfactory, but surface temperatures had risen to 80°C. Figure 11.11 shows the transition of the site into methanogenesis with the accom-

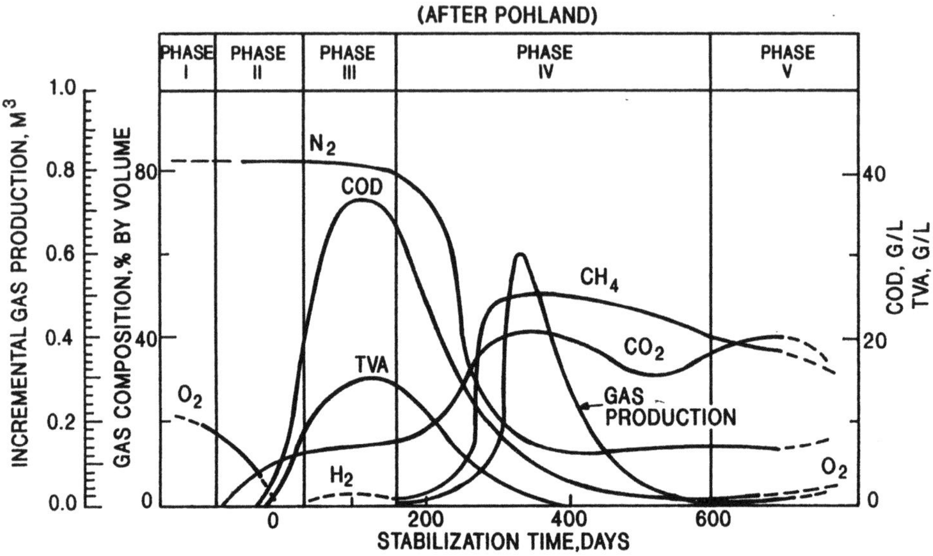

Fig. 11.12. Phases of landfill gas generation (Pohland 1987)

panying rise in temperature (CEMS 1995). The various stages of landfill gas generation are shown in Fig. 11.12 (Pohland 1987). This phased generation of gases, assuming optimum conditions exist within the site, contrasts with the experience of generation of methane at the Craigmuschat site after a number of years of waste disposal.

In order to study the process of gas and leachate generation of a landfill, an experimental site was set up at Mid Auchencarroch (Wingfield-Hayes et al. 1995). The purpose of this experimental landfill site was to study the gas generation from different waste streams and in particular cases from mono-streams such as timber. The results for the initial research indicated vast differences in the rate at which gas generation occurred with high variation in the concentrations of gas. Detailed assessment of the trace gases being emitted from the site are also being studied to determine how the primary waste used in landfill sites breaks down into hazardous or toxic by-products. Recent research in California (Webster 1988) clearly indicates as shown in Fig. 11.13 that certain primary wastes can break down into hazardous trace gases. The most important of these is lignin, which can break down into benzene, toluene and phenol. Lignin is a primary constituent of timber and paper, the rapid increase on timber and paper going to landfill may pose a problem for trace gases in the future. The purpose of the research is to determine the effect of removing these primary waste streams, such as paper and that effect on the final gas and leachate quality generated within the landfill site.

SUMMARY

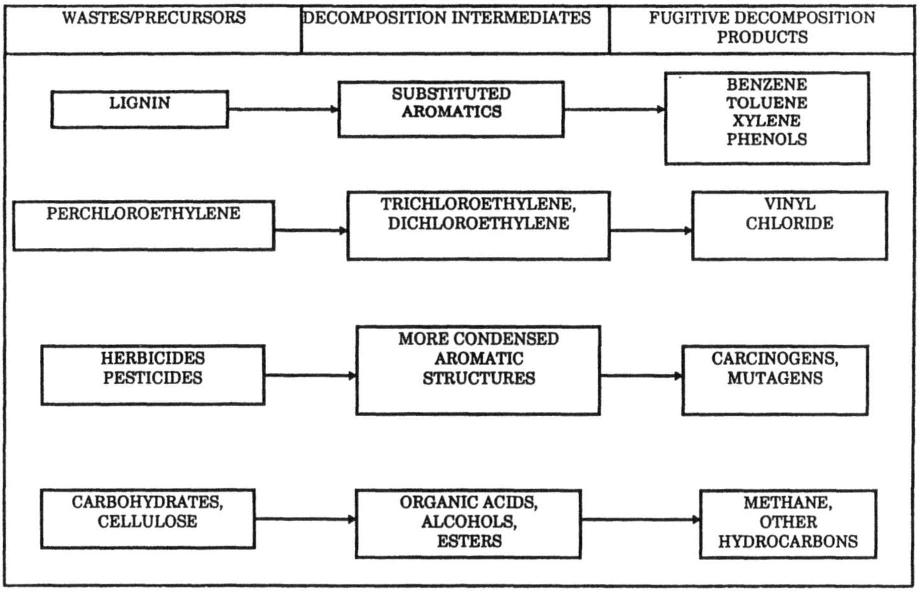

Fig. 11.13. Reactions leading to the generation of toxics from solid waste (Webster 1988)

Landfill operational management is, therefore, of key importance to understanding the waste mix and the processes and how the landfill site is filled and to monitoring following closure of the landfill site. This requires monitoring of the waste stream, together with monitoring of emissions from the landfill site as it is filled and in the after-care phase. Table 11.8 (Draft EEC Directive 1995) shows the monitoring which may be recommended by the European Directive on landfill for the leachate and gas emissions. In addition, meteorological data is also shown in Table 11.8 together with the groundwater monitoring and topographic surveys. The amount of data generated by the monitoring during operational after-care phases of landfill will lead to a burst in the information base and hopefully will allow better understanding of landfill process as it impacts on the environment. However, to handle this information explosion, it requires modern information management systems such as "EnviSage" (Centre for Environmental Management Studies 1995b) which allow the structured management of the information flow and the interpretation of the data through graphical interpretation and statistical analysis. Figure 11.14 shows a typical image for a landfill site with monitoring points and Fig. 11.15 shows a typical display of the information.

Table 11.8. Monitoring recommended for leachate and gas emissions (Draft EEC Directive 1995)

		Operating phase	Aftercare phase[3]
2.1	Leachate volume	monthly[1][3]	every 6 months
2.2	Leachate compositicn[2]	quarterly[3]	every 6 months
2.3	Volume and composition of surface water[7]	quarterly[3]	every 6 months
2.4	Potential gas emissions and atmospheric pressure[4] (CH_4,CO_2,O_2,H_2S,H_2 etc.)	monthly[3][5]	every 6 months[6]

(1) The frequency of sampling could be adapted on the basis of the morphology of the landfill waste (in tumulus, buried, etc.). This has to be specified in the permit.
(2) The parameters to be measured and the substances to be analyzed vary according to the composition of the waste deposited; they must be laid down in the permit document and reflect the leaching characteristics of the wastes.
(3) If the evaluation of data indicates that longer intervals are equally effective, they may be adapted. For leachates, conductivity must always be measured at least once a year.
(4) These measurements are relevant mainly to landfills receiving large quantities (>25% w/w) of organic waste.
(5) CH_4, CO_2, O_2 regularly, other gases as required, according to the composition of the waste deposited, with a view to reflecting its leaching properties.
(6) Efficiency of the gas extraction system must be checked regularly.
(7) On the basis of the characteristics of the landfill site, the competent authority may determine that these measurements are not required, and will report accordingly in the way foreseen in Article 16 of the Directive.

2.1 and 2.2 apply only where leachate collection takes place (see Annex I(2))

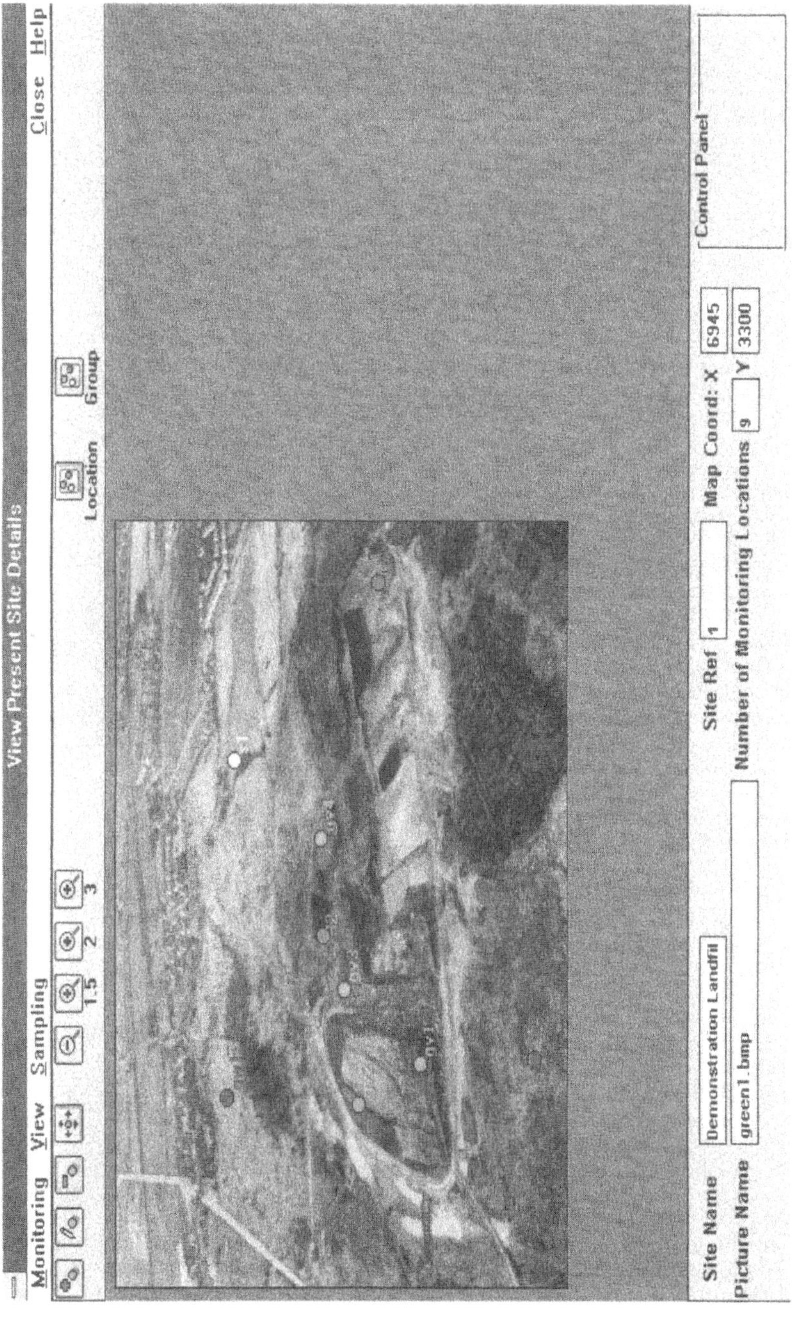

Fig. 11.14. EnviSage – Typical image of landfill site

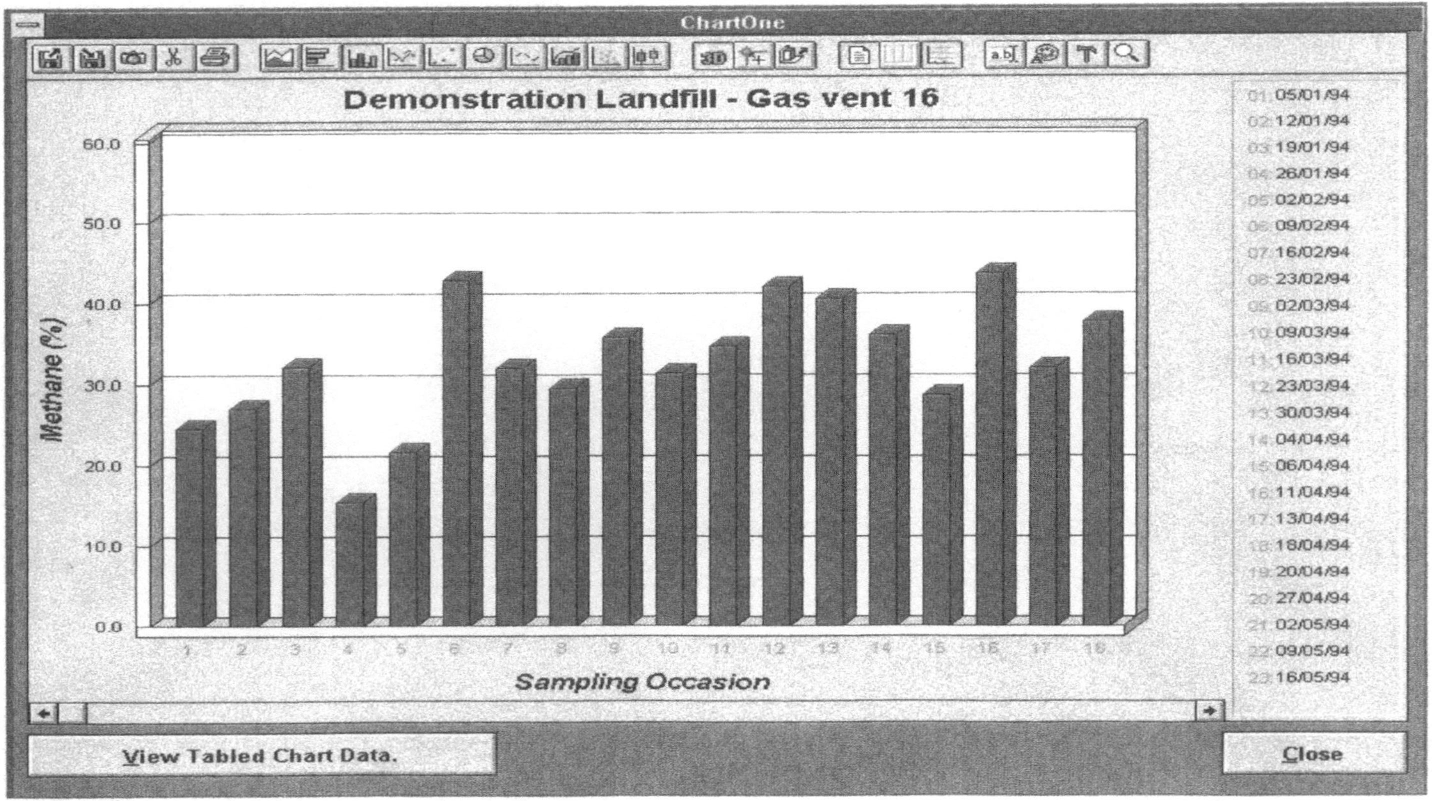

Fig. 11.15. EnviSage – Typical display of information

11.7 Discussion

The present state of landfilling knowledge is simply no more than a black box with both the inputs and the outputs from the landfill system only understood in qualitative terms with very limited quantitative understanding. The final storage quality of a landfill defined as near zero pollution potential of solids and pollution streams of liquids cannot be reached during the next centuries (Förstner 1988). With the increasing trend towards sophisticated systems for landfill design, operation and management, there remains a continuing failure of such systems. Landfill sites are presently being designed as if they were part of the geological formation but without due consideration of the extremely long time periods over which landfills will decay. The rapid change in legislation reflects the politician's will to avoid short-term problems. Unfortunately the short-term legislation e.g. the proposed landfill tax of the UK government will often act as a catalyst for future failure of waste disposal systems by forcing change in the various waste streams which are not sustainable in the long term.

The way forward for future landfill practice is the increasing use of geochemical engineering methods to ensure that the natural resources available at the landfill site are reprocessed in such a way as to reduce the risk of polluting emissions impacting on the environment.

References

Boels D, Fleming G (1993) Chemical time bombs from landfills: appraisal and modelling. Land degradation and rehabilitation, Vol 4, John Wiley & Sons Ltd

Centre for Environmental Management Studies (1995a) Results compiled by CEMS from material supplied by Department of Environmental Health, Inverclyde District Council

Centre for Environmental Management Studies (1995b) EnviSage, University of Strathclyde, Glasgow

Christianson T H et al. (1994) Attentuation of landfill leachate – pollutants in aquifers, In: Critical reviews in environmental science and technology 24(2):119–202.CRC Press Inc

Crawford J, Smith P (1985) Landfill technology, Butterworths, London

EEC Draft Directive (1995) 4103/95. General Secretariat of European Union, Brussels

Ehrig H J (1983) Quality and quantity of sanitary landfill leachate. Waste Management Resource, 1, 53

Ehrig H J (1988) Water and element balances of landfills. In: The landfill, Baccini P (ed) Lecture Notes in Earth Sciences, Vol 20. Springer, Berlin Heidelberg New York

Environmental Protection Act (1990) HMSO, London

Fleming G (1989) Investigation into leachate emission from Brackletter landfill, Spean Bridge. Report to Locaber District Council, Fort William

Fleming G (1991) A review of the management of waste disposal at Greenoakhill landfill site, Vols I and II, WEMU, University of Strathclyde, Glasgow

Fleming G (1992) Landfills and contaminated land: time delayed responses. University of Strathclyde, Glasgow

Fleming G and Macdonald M (1994) Gas production levels and leachate from landfill sites of differing waste disposal practices in Central Scotland. University of Strathclyde, Glasgow

Förstner U (1988) Geochemical processes in landfills,. In: Sounder Bhaltacharj S, Friedman GM, Horst J, Neugebauer and Sieldman A, Earth Sciences. Technical University, Hamburg

Giroud JP, Bonaparte R (1989) Leakage through liners constructed with geomembranes – Part I, Geomembrane liners, Part II, Composite liners. geotextiles and geomembranes, 8:27–68; 71–112

Harrison (1976) Institute of Geological Sciences

HMSO (1989) The control of landfill gas. Waste Management paper no 27, London

National Geographic (1991) Once and future landfills. Article by WL Rathje, pp 116–134

Nicholson R V, Cherry J A, Reardon E J (1983) Migration of contaminants in groundwater at a landfill: a case study, VI Hydrogeochemistry, J Hydrol, 63, 131

Noble G (1992) Siting landfills and other LULUs, Technomic Publishing Company Inc, USA

Peters T (1993) Chemical and physical changes in the subsoil of three waste landfills. Waste Management and Research 11, 17–25

Pohland FG (1987) Proceedings of third international symposium on anerobic digestion,. Boston, USA

Rowe RK, et al. (1995) Clayey barrier systems for waste disposal facilities. E & F N Spon, London

Times (1995) The Times Newspaper, 5th October 1995, p 32, London

Webster I (1988) Municipal solid waste landfills in Superfund. Address to National Association of Manufacturers, California

Wingfield-Hayes C et al. (1995) Developing appropriate technology for small scale landfill, University of Strathclyde, Glasgow

Subject Index

Springer-Verlag and the Environment

MIX
Papier aus verantwortungsvollen Quellen
Paper from responsible sources
FSC® C105338

If you have any concerns about our products,
you can contact us on
ProductSafety@springernature.com

In case Publisher is established outside the EU,
the EU authorized representative is:
Springer Nature Customer Service Center GmbH
Europaplatz 3, 69115 Heidelberg, Germany

Printed by Libri Plureos GmbH
in Hamburg, Germany